INCREASING MOTORCYCLE CONSPICUITY

Human Factors of Simulation and Assessment

Series Editors

Michael Lenné
Monash University Accident Research Centre, Melbourne, Australia
Mark Young
School of Engineering and Design, Brunel University, London, UK

Ongoing advances in lower-cost technologies are supporting a substantive growth worldwide in the use of simulation and naturalistic performance assessment methods for research, training and operational purposes in domains such as road, rail, aviation, mining and healthcare. However, this has not been accompanied by a similar growth in the expertise required to develop and use such systems for evaluating human performance. Whether for research or practitioner purposes, many of the challenges in assessing operator performance, both using simulation and in natural environments, are common. What performance measures should be used, what technology can support the collection of these measures across the different designs, how can other methods and performance measures be integrated to complement objective data, how should behaviours be coded and the performance standards measured and defined? How can these approaches be used to support product development and training, and how can performance within these complex systems be validated? This series addresses a shortfall in knowledge and expertise by providing a unique and dedicated forum for researchers and experienced users of simulation and field-based assessment methods to share practical experiences and knowledge in sufficient depth to facilitate delivery of practical guidance.

Increasing Motorcycle Conspicuity

Design and Assessment of Interventions
to Enhance Rider Safety

Edited by

LARS RÖßGER
University of Technology, Dresden, Germany
MICHAEL G. LENNÉ
Monash University Accident Research Centre, Melbourne, Australia
and
GEOFF UNDERWOOD
University of Nottingham, UK

CRC Press
Taylor & Francis Group
Boca Raton London New York

CRC Press is an imprint of the
Taylor & Francis Group, an **informa** business

CRC Press
Taylor & Francis Group
6000 Broken Sound Parkway NW, Suite 300
Boca Raton, FL 33487-2742

First issued in paperback 2017

© 2015 by Lars Rößger, Michael G. Lenné and Geoff Underwood
CRC Press is an imprint of Taylor & Francis Group, an Informa business

No claim to original U.S. Government works

ISBN-13: 978-1-4724-1112-9 (hbk)
ISBN-13: 978-1-138-74764-7 (pbk)

Visit the Taylor & Francis Web site at
http://www.taylorandfrancis.com

and the CRC Press Web site at
http://www.crcpress.com

Contents

List of Figures

List of Tables

List of Contributors

Vanessa Beanland, Research School of Psychology, Australian National University, Canberra, Australia

Viola Cavallo, French Institute of Science and Technology for Transport, Development and Networks (IFSTTAR), Laboratory for Road Operations, Perception, Simulators and Simulations (LEPSIS), France

Zoi Christoforou, Ecole des Ponts et Chaussées, 6&8 avenue Blaise Pascal – Cité Descartes, Champs-sur-Marne, France

Stéphane Espié, French Institute of Science and Technology for Transport, Development and Networks (IFSTTAR), Laboratory for Road Operations, Perception, Simulators and Simulations (LEPSIS), France

Pnina Gershon, Carnegie Mellon University, Department of Psychology, Human Factors Engineering & Driving Safety, Pittsburgh, United States of America

John Golias, Department of Transportation Planning and Engineering, School of Civil Engineering, National Technical University of Athens, Greece

Katherine Humphrey, University of Nottingham, School of Psychology, United Kingdom

Jens Krzywinski, University of Technology Dresden, Faculty of Mechanical Science and Engineering, Industrial Design Engineering, Germany

Michael G. Lenné, Monash University Accident Research Centre, Monash University, Victoria, Australia

Eve Mitsopoulos-Rubens, Monash University Accident Research Centre, Monash University, Victoria, Australia

Frank Mühlbauer, University of Technology Dresden, Faculty of Mechanical Science and Engineering, Industrial Design Engineering, Germany

Maria Pinto, French Institute of Science and Technology for Transport, Development and Networks (IFSTTAR), Laboratory for Road Operations, Perception, Simulators and Simulations (LEPSIS), France

Joceline Rogé, Laboratory of Ergonomics and Cognitive Sciences applied to Transport (LESCOT) – Transport, Health, Safety Department, French Institute of Science and Technology for Transport, Development and Networks (IFSTTAR), France

Lars Rößger, University of Technology Dresden, Faculty of Traffic Science, Traffic and Transportation Psychology, Germany

Peter Saleh, AIT Austrian Institute of Technology GmbH, Mobility Department, Transportation Infrastructure Technologies, Austria

Bernhard Schlag, University of Technology Dresden, Faculty of Traffic Science, Traffic and Transportation Psychology, Germany

David Shinar, Ben Gurion University of the Negev, Israel

Geoff Underwood, University of Nottingham, School of Psychology, United Kingdom

Editha van Loon, University of Nottingham, Division of Psychiatry, United Kingdom

Fabrice Vienne, Laboratory for Road Operations, Perception, Simulators and Simulations (LEPSIS) – Component and Systems Department, French Institute of Science and Technology for Transport, Development and Networks (IFSTTAR), France

George Yannis, Department of Transportation Planning and Engineering, School of Civil Engineering, National Technical University of Athens, Greece

Foreword

Stéphane Espié

'Sorry mate I didn't see you' is one of the most common explanations given by drivers when they cut-across motorcyclists at intersections. 'I didn't see him' is invoked by pedestrians being hit by motorcyclists, for instance at pedestrian crossings.

The difficulties to properly detect the arrival of powered two-wheeled vehicles' (PTW) and to evaluate their approaching speed are clearly concerns for road safety. The problem is not only perceptual but also cognitive. The relatively low number of PTW in the traffic can partially explain a bad understanding, thus anticipation, of their specific behaviours.

To improve their detectability, and in many countries, PTW have for years to use their daytime-running lights (DRL) in the day time. The recent broadening of the use of car lights during the day, with various marketing-based signatures, may sometimes be a problem for the detection of PTW due to an increase of visual noise.

Several approaches may be proposed to increase the PTW/rider conspicuity, by enlarging the visible shape of the vehicle and/or of the rider. However many proposed solutions where not supported by scientific evidence …

The 2BESafe project aimed at conducting a broad set of scientific research to better understand the motives that underlie PTWs' over-involvement in road accidentology. The 2BESafe program was structured into six research work-packages that included: fundamental research on crash causes and human error (WP1); the world's first Pan-European naturalistic driving study involving instrumented PTWs (WP2); an experimental research on motorcycle rider risk awareness and perception (WP3); the development of research tools to support the 2BESafe human factors and behavioural research program (WP4); a large-scale research program on the factors that underlie drivers' failure to see PTWs and their riders (WP5); and the development of practical countermeasures for enhancing PTW rider safety deriving from all these activities (WP6) (see www.2besafe.eu for more information).

Within WP5, particular attention was given to the PTW conspicuity issue, and several studies have been conducted aimed at paving the way of future improvements in terms of vehicle and/or rider visibility, some of them requiring some changes in the regulation. This work was achieved within the WP5.2 task that involved Dresden Technical University (Germany), Ben Gurion University

(Israel), INRETS/LPC (France), INRETS/MSIS (France), Monash University (Australia) and Nottingham University (United Kingdom).

This book synthesizes this late research field, and I hope it will help researchers, practitioners and stakeholders to propose, in the near future, relevant improvements for road safety regarding PTWs.

PART I
Setting the Stage: Motorcycle Safety and Conspicuity

Chapter 1

PTW Crashes and the Role of Perception

Zoi Christoforou, George Yannis, John Golias and Peter Saleh

Introduction

Powered two-wheelers (PTWs) are a vulnerable class of road users with *increased accident frequency and severity* (Vlahogianni et al., 2012). In the early 1990s, motorcycle death-rate-per-mile-travelled was estimated to be 22 times the death rate for passenger cars (Preusser et al., 1995). In 2007, US motorcycle riders had a 34-fold higher risk of death in a crash than people driving other types of motor vehicles (NHTSA, 2007). In 2008, European motorcyclists represented 17 per cent of road fatalities while only accounting for 2 per cent of road users (IRTAD, 2009). In Greece this figure is as high as 33 per cent (IRTAD, 2013) while in Singapore it reaches 49 per cent with more than two motorcyclists being killed every week (Haque et al., 2012). Higher crash risk is associated to the fact that driver- and rider-related factors are much more prevalent in PTW accidents compared to vehicle- and environment-related factors. In particular, there exists a clear over-representation of inappropriate perception in PTW crashes (Van Elslande et al., 2013). One often discussed reason for perception failures is that PTW are less conspicuous than other motorized road users (Rößger et al., 2012). Consequently, gap acceptance is often inadequate due to the size-arrival illusion (Horswill et al., 2005); the latter refers to small objects being perceived to arrive later than larger ones. Besides conspicuity, car drivers seem to encounter difficulties in understanding PTWs' manoeuvres and, thus, fail to foresee PTWs' behaviour; foresight is the result of the combination of circumstantial data and permanent knowledge and beliefs (Ragot-Court et al., 2012).

Conspicuity can be examined from two different angles; namely the *sensory and the cognitive perspectives*. Sensory conspicuity is the visual distinction of an object due to its physical characteristics (Wulf et al., 1989). It refers to the extent to which an object can be distinguished from its environment because of its characteristics: angular size, eccentricity, brightness against the background, colour and so on. It reflects an object's ability to attract visual attention and to be precisely located as a result of its physical properties (Rogé et al., 2012). The size and vehicle dynamics of PTWs are such that they have lower sensory conspicuity (Gershon and Shinar, 2013). Cognitive conspicuity depends on the distinction of an object based on the observer's experiences and interests (Wulf et al., 1989). It is linked to the fact that an observer's focus of attention is strongly influenced by his/her expectations, objectives and knowledge (Rogé et al., 2012). Interestingly, inappropriate expectations may be even more important in accident causation than the motorcyclist's physical properties (Hole et al., 1996). PTWs show lower cognitive conspicuity as they account for relatively few vehicle miles travelled compared to automobiles, especially in Western countries (Gershon et al., 2012). Furthermore, not all car drivers have previous PTW riding experience. Helman et al. (2012) distinguishes further among:

- visibility: the extent to which an object stands out from its surroundings when observers are *aware* of its location;
- search conspicuity: the extent to which an object stands out from its surroundings when observers are *searching* for it within a scene;
- attention conspicuity: the extent to which an object stands out from its surroundings when observers are viewing the scene, but *not searching* deliberately for the object.

Vision is the predominant sensory modality used when driving (Crundall, 2011). Consequently, *conspicuity is an important* issue to all road users: be it cyclists (see, for example, Lacherez et al., 2013; Madsen et al., 2013); pedestrians (see for example, Tyrrell et al., 2004); or car drivers (see, for example, Alferdinck, 2004; Berg et al., 2007). However, evidence shows that vulnerable road users tend to underrate the role of visibility factors and conspicuity benefits (Lacherez et al., 2013) while overestimating their own conspicuity (Wood et al., 2013). Comparisons between bicycle and motorcycle crashes suggest that the majority of both crash types occur at intersections and are due to conspicuity issues (Haworth and Debnath, 2013). Nevertheless, PTWs seem to be more concerned due to a combination of factors including high speeds and acceleration rates (if compared to cyclists and pedestrians) and small size (if compared to other motorized road users).

Indeed, PTW conspicuity has been long been recognized as a critical PTW *crash contributory* factor. In 1975, the Greater London Road Safety Unit identifies a certain PTW over-representation in accidents. Detailed analysis of crash data followed. Results indicated that a major contributory factor was the failure of other drivers to observe PTWs in the general street scene (Lalani and Holden, 1978). Riders were then encouraged to wear bright clothing, preferably of fluorescent material and to switch on headlights during the daytime. A lot of research has been undertaken since 1975 on the so-called 'PTW conspicuity hypothesis'. Accident investigations have been carried out in many countries and report that between half and three-quarters of motorcycle accidents involve collision with another vehicle (Huang and Preston, 2004). Markedly, most right-of-way (ROW) accidents involving PTWs are attributed to conspicuity (Pai et al., 2009) while drivers of other vehicles are at fault in the majority of two-unit motorcycle crashes (Haworth and Debnath, 2013).

In view of the above, this chapter's objective is to examine the main determinants of riders' accident risk which are related to conspicuity issues. We perform a literature review in order to explore the role of conspicuity in PTW crash occurrences. English-language publications were selected for relevance through a comprehensive search of major databases (see Table 1.1). The key words used in the search were: 'conspicuity' and 'motorcycles'. To be included, papers were assessed against additional criteria; mainly relevance and publication date. The latter was thought to be critical as earlier literature reviews do exist. In the following section, we briefly discuss the effectiveness of conspicuity interventions. A detailed overview can be found in Chapters 4–10. In the third section, we review PTW accident risks and severity outcomes. Finally, we summarize findings and provide conclusions.

Conspicuity Interventions and Contextual Factors

PTW conspicuity risk can be defined as an increased probability of 'low' conspicuity. As many previous studies show (Helman et al., 2012; Lin and Kraus, 2009; Pai, 2011; Wulf et al., 1989) the conspicuity level is changing, relative, and largely dependent upon contextual factors. PTW conspicuity may be related to the motorcycle, to other vehicles, to the riders themselves, to other drivers, to the road environment or to any combination of those factors. Moreover, they may be associated to exogenous or endogenous, modifiable or non-modifiable factors. For example, riders can use daytime running lights to decrease their probability of collision with another vehicle (Saleh et al., 2010), but they do not control ambient traffic conditions. Also, frontal, lateral and rear motorcycle sensory conspicuity may differ significantly. Most importantly, conspicuity is not constant but changes with the time of day, the weather conditions, the urban environment, the presence or absence of other road users. A negative or neutral element, such as a dark jacket in night-time conditions, may have a positive impact in daytime. Inversely, a daytime conspicuity intervention may prove to have a negative impact during night-time. Therefore, it is difficult (if not impossible) to establish a rigorous taxonomy of conspicuity risks and to assess their impact under all possible circumstances. The related literature considers different road environments (rural vs. urban, intersections, light vs. heavy traffic), varying lighting conditions and driver attributes and has mainly focused on the following type of measures:

- Vehicle lights (Cavallo and Pinto, 2012; Farmer and Williams, 2002; Jenness et al., 2011; Hole et al., 1996; Janoff and Cassel, 1973; Lenné and Mitsopoulos-Rubens, 2011; Muller, 1982; Perlot and Prower, 2003; Rumar, 1980; Smither and Torrez, 2010; Thomson, 1980; Umar et al., 1996; Yuan, 2000; Zador, 1985).
- Rider clothing and motorcycle colour (Burg and' Beers, 1978; Gershon et al., 2012; Hole et al., 1996; Olson et al., 1981; Smither and Torrez, 2010; Watts, 1980; Williams and Hoffmann, 1979).

- Rider experience (ACEM, 2004; Crundall et al., 2012; Crundall et al., 2008; Mitsopoulos-Rubens and Lenné, 2012).

Riders' Accident Risk

PTWs are believed to have a higher risk of getting involved in accidents compared to other vehicle drivers. If involved in accidents, PTWs are also more likely to experience severe injuries. In New Zealand, for instance, motorcyclists represent 13 per cent of deaths and 9 per cent of road injuries while motorcycles represent only 3.5 per cent of registered vehicles (Helman et al., 2012). This over-representation is even greater if considering the lower mileage of motorcycles: they undertake around 0.5 per cent of travel time or trip legs (Walton et al., 2013). Furthermore, the cause of the majority of PTW accidents is human error and the most frequent human error is a failure to see the PTW within the traffic environment, due to lack of driver attention, temporary view obstructions or PTW low conspicuity (ACEM, 2004). Zador (1985) relates conspicuity to single-vehicle accidents. He claims that one-fifth of PTW single-vehicle accidents result from riders trying to avoid other vehicles. However, low conspicuity is primarily associated with car-versus-motorcycle (CVM) collisions. Inadequate motorcycle visibility is an associated factor in 64.5 per cent of CVM collisions and it is the sole identifiable cause of 21 per cent of collisions (Williams and Hoffmann, 1979).

In CVM collisions, car drivers are mostly at fault: the most common motorcycle crash type is when an automobile manoeuvres into the path of an oncoming motorcycle at an intersection which involves a motorist infringing upon the motorcycle's right-of-way (ROW) (Helman et al., 2012; Pai et al., 2009; Wulf et al., 1989). The motorcycle's ROW is more likely to be violated at unsignalized T-junctions (Pai and Saleh, 2008), non-built-up roads and in poor light conditions (Pai et al., 2009). In an early effort, Fulton et al. (1980) reported that about 67 per cent of near-misses and motorcycle accidents were due to another driver failing to detect the oncoming motorcyclist before emerging from a side turning or before turning across the motorcyclist's path. Preusser et al. (1995) explored a US database of 2,074 crashes fatal to the motorcycle rider and conclude that approximately one-quarter of total crashes are due to some other vehicle failing to grant the ROW and moving into the path of the motorcycle. ROW violations are involved in 40 per cent of all CVM crashes in Great Britain (Clarke et al., 2007) and 64 per cent of CVM crashes (Walton, 2010) in New Zealand. The frequency of this crash pattern is such that PTW ROW violation by another vehicle has become representative of both CVM collisions and conspicuity-related crashes. Umar et al. (1996) define conspicuity-related motorcycle accidents as 'all accidents involving motorcycles travelling straight or turning onto a ROW and colliding with pedestrians and other vehicles'. However, other crash types (single-vehicle accidents for example) and different pre-crash manoeuvres (overtaking for example) may be also related to low PTW conspicuity. Inversely, PTW ROW violations may be due to reasons other than low conspicuity. Sometimes drivers do not look at all when pulling out of a junction; this is not a conspicuity issue (Helman et al., 2012). Nevertheless, this information is available only in laboratory experiments. Most of the accidents (65 per cent) collected in straight sections were motor vehicle collisions between a passenger car and a PTW. Almost half of accidents occurred in darkness, suggesting a problem of sensory conspicuity (Spanish investigation within the project 2BESafe; Saleh et al., 2010).

First and foremost, car drivers violate motorcycle ROW because they 'look but fail to see' (LBFS). LBFS accidents happen when a driver pulls into the path of an oncoming motorcyclist and claims not to have seen him/ her approaching (Herslund and Jorgensen, 2003). LBFS accidents mostly occur in daytime. Indeed, daytime PTW conspicuity is lower as, during night-time, headlights provide a strong contrast to the lighting environment (Wulf et al., 1989). Secondarily, car drivers violate motorcycle ROW because they fail to correctly judge the path or speed of the PTW (Gould et al., 2012a). CVM collisions then occur as a result of drivers accepting an inadequate gap among conflicting traffic (Pai et al., 2009). Experimental evidence proves that drivers make more accurate judgements regarding the approaching speed of cars than the speed of motorcycles, especially in night-time conditions (Gould et al., 2012a). Motorcyclists often experience reduced visibility when wearing glasses, visors or wind shields (NPRA, 2004).

Lin and Kraus (2009) classify conspicuity in a Haddon's matrix as a pre-event risk factor related to human, vehicle and environmental crash aspects. Besides the three interventions discussed previously, the following factors seem to be influential:

a. *Human factors*
   Age and gender have an impact on identification and reaction times or even on the effectiveness of conspicuity aids (Smither and Torrez, 2010). Elderly and female motorists appear to be over-represented in gap-acceptance crashes (Pai et al., 2008). Magazzù et al. (2006) suggest that motorcycle conspicuity is lower among older car drivers. Clarke et al. (2007) provide evidence that older and experienced drivers seem to have more problems detecting approaching motorcycles particularly at T-junctions. Injuries to riders are greatest in angle oblique collisions with elderly motorists while teenage motorists seem to predispose riders to a greater injury risk in angle perpendicular crashes while (Pai, 2009). Furthermore, some authors attribute car drivers' failure at junctions to the higher workload during turning manoeuvres at intersections (Hancock et al., 1990) or even to their negative view towards motorcyclists (Crundall et al., 2008). It should be noted that many human factors that are critical to road safety (fatigue, alcohol impairment, drug use and so on) have not been examined under the conspicuity hypothesis.

b. *Vehicle speed and distance*

PTW distance from the viewer is not only a contributing conspicuity factor but also influences the effectiveness of different aids in increasing conspicuity (Gershon and Shinar, 2013). The possible influence of speed on low motorcycle conspicuity has been suggested by a number of authors (see Kim and Boski, 2001; Williams and Hoffmann, 1979). Clabaux et al. (2012) examined the effect of motorcyclists' speed on their involvement in LBFS accidents in France. The authors performed a kinematic reconstruction of 44 accident cases occurring in both urban and rural environments. Results indicate that in urban environments the approach speed of motorcyclists involved in LBFS accidents is significantly higher than in other accidents at intersections. In rural environments, the speed difference was not found to be significant.

c. *Road environment*

Motorcycles' ROW is more likely to be violated on non-built-up roads (Pai et al., 2008). Nevertheless, evidence shows that PTW crashes mostly occur in urban areas while passenger cars are the most frequent collision partners (ACEM, 2004). In the ACEM study (2004), over half of PTW crashes took place at intersections while 90 per cent of all PTW accidents occurred in light to moderate traffic conditions. Poor visibility conditions (horizontal curvature, vertical curvature, darkness) are responsible for increased motorcycle injury severity (Savolainen and Mannering, 2007). Poor sight-line visibility and rider/bike conspicuity are likely to contribute to motorcycle accidents at intersections (NPRA, 2004). Moreover, riding in darkness without street lighting was related to severe motorcyclists' injury (De Lapparent, 2006; Pai and Saleh, 2007, 2008). Motorcyclists are found to be more vulnerable during night time at both intersections and expressways (Haque et al., 2009). Injuries resulting from early morning riding, in general, appear to be the most severe, especially in junctions controlled by stop, and give-way signs and markings (Pai and Saleh, 2007).

Haque et al. (2012) explored motorcycle crash occurrences in Singapore where motorcycles account for 16.3 per cent of motorized vehicle fleets. The authors specified a log-linear model over a database, including a total of 13,568 occurring on expressways, at intersections or away from intersections. Night-time influence was found to increase crash risk particularly during merging and diverging manoeuvres on expressways, and turning manoeuvres at intersections. The authors suggest that this is due to night-time conspicuity. Of course, conspicuity explains to an extent the latter but other factors may come into play as well: lower traffic volumes and higher speeds, more sensation-seeking and risk-taking behaviours and so on. Intersections (poor sight-line visibility and rider/bike conspicuity are likely to contribute to motorcycle accidents at intersections). Analyses of Spanish PTW crash data show that the most frequent type of intersection where accidents occurred is a roundabout (7 out of 8) in interurban areas. Most of the accidents collected in these junctions occurred without daylight conditions so it could be suggested that kerbs should be painted with the aim of raising their conspicuity (Saleh et al., 2010).

Overall, accident studies and *post-hoc* crash investigations establish only indirect links between crash outcomes and conspicuity factors and interventions. The difficulty in directly associating conspicuity interventions to safety outcomes starts from the very definition of conspicuity-related motorcycle accidents that remains rather unclear. A second major barrier to establishing this link is that conspicuity-related factors cannot be collected from conventional (national) crash databases (Shaheed et al., 2012). In the absence of relevant data, researchers mainly perform before-after evaluations or longitudinal studies comparing crash data with and without the treatment. In all these cases, the presence of bias – due to site particularities or other reasons – cannot be excluded. A third methodological problem consists in comparing among subsets of crash dataset: single- vs. multi-vehicle motorcycle accidents, daytime vs. night-time motorcycle accidents and so on. Such comparisons juxtapose crashes with clearly different causes. Comparisons between groups of crashes with common causes (for example, car drivers' failure to detect a car) would be more appropriate. Besides, empirical evidence shows that CVMs are not that different from CVCs. Cercarelli et al. (1992) investigated 500 CVM crashes and compared them to over 3,000 CVC crashes. The analysis did not identify any consistent pattern between crash-type and lighting conditions. Walton et al. (2013) performed a case-control study between CVC and CVM crashes in New Zealand. This analysis again showed that CVM crashes are not easily distinguished from CVC crashes as they follow similar patterns.

We identified only two recent studies establishing empirical causal links between conspicuity and motorcycle accident risk. In 2004, the ACEM funded a comprehensive Motorcycle Accident In-Depth Study (MAIDS) project that covered five European countries: France, Germany, Italy, Spain and the Netherlands. The authors compared 921 motorcycle accident cases with 923 controls and offered very interesting insights on conspicuity contributory factors. White PTWs were found to be over-represented in crash occurrences. Dark PTW rider clothing decreased conspicuity in 13 per cent of all accidents. Wells et al. (2004) designed an innovative population-based case-control study in New Zealand. The authors interviewed 463 motorcycle riders (cases) involved in car-motorcycle crashes resulting to the motorcyclist's injury or death. In the latter case, a proxy respondent was interviewed instead. In addition, 1,233 motorcycle riders were randomly recruited and interviewed (controls). Statistical analysis of responses revealed that injury crashes mainly occurred in urban zones with 50km/h speed limit, during the day and in fine weather. Riders wearing reflective and fluorescent clothing had a 37 per cent lower risk. The use of a white helmet was associated with a 24 per cent lower risk compared to a black helmet. DRL was found to be associated with a 19 per cent lower risk of involvement in injury crashes. No association was found between risk and the frontal colour of rider's clothing or motorcycle.

Turning to accident severity, head injuries are the most frequent in fatal motorcycle crashes, accounting for about 50 per cent of all motorcycle deaths (Kraus, 1989). Chest and abdominal injuries are the second most frequent cause comprising from 7 per cent to 25 per cent of deaths (Lin and Kraus, 2009). If all injury outcomes are considered, lower extremity injuries come first, followed by upper extremity injuries (ACEM, 2004). As in the case of accident involvement, few of the studies linked conspicuity to severity outcomes in a consistent way. Pai (2009) examined how motorists' failure to give way affects motorcyclist injury severity at T-junctions in UK. Binary logit estimation results revealed that injuries are greater when a travelling-straight motorcycle on the main road crashes into a right-turn car from the minor road. Shaheed et al. (2011) used the Iowa crash database to explore differences in motorcycle severities in daylight and dark conditions. It seems that more severe car-motorcycle crashes are likely to occur in dark conditions. Furthermore, rear-end crashes caused by a 'non-motorcycle' vehicle hitting a motorcycle are less likely to result in severe injuries. On the contrary, a higher likelihood of major injuries was found in angle crashes with the 'non-motorcycle' vehicle turning left and the motorcycle moving straight.

Conclusions

The objective of this chapter was to examine the main determinants of riders' accident risk that are related to conspicuity by means of a literature review. We tried to establish a link between conspicuity risks/enhancements and PTW crash risk. Conspicuity risks are not to be confused with accident risks; they are two separate sets of risk factors. The conspicuity hypothesis testing consists in proving that an intersection exists between the two and in defining its elements. In that sense, the conspicuity hypothesis is not rejected for a treatment that is found to significantly enhance conspicuity *and* to significantly decrease PTW accident risk.

Empirical evidence shows that most PTW conspicuity crashes are related to PTW ROW violation by another vehicle and result to a collision between the two. These accidents seem to be most probable at unsignalized intersections, non-built-up roads and in poor light conditions. The most frequent collision partners are passenger cars. It should be noted though that other crash patterns exist and that not all PTW ROW violations are due to conspicuity issues. Unfortunately, common accident *records do not include specific information* on conspicuity aids at use at the moment of the accident. Also, laboratory experiments may simulate the driving environment but not the real-world crash conditions. In the absence of data, most researchers have either investigated PTW crash data through observational studies or explored the effectiveness of conspicuity treatments by means of laboratory experiments. In both cases, it is impossible to infer empirical links between conspicuity and accident involvement. The added value of laboratory studies is that they allow researchers to better understand the perceptual and behavioural issues that *may contribute* to accident risk as they aim to measure the conditions

Table 1.1 Overview about findings from recent studies on the link between conspicuity and crash statistics

Author	Objective	Data	Method	Methodology	Conspicuity configuration	Findings
ACEM (2004)	To better understand the nature and causes of PTW accidents	921 accidents (Europe, 1999–2000); 923 controls	In-depth investigation	Case-control study	Motorcycle colour Lighting conditions Headlamps on/off Motorcyclist clothing colour	White PTWs are over-represented in crash occurrences. Car drivers holding a PTW licence are less likely to commit perception failures Dark PTW rider clothing decreased conspicuity in 13% of crashes
Clabaux et al. (2012)	To examine the effects of motorcyclists' speed on LBFS accidents	44 LBFS accident cases (France)	In-depth investigation	Kinematic reconstructtion	Motorcycle speed	In urban environments, the motorcycles' speed involved in LBFS accidents is significantly higher In rural environments, the speed difference is not significant
Clarke et al. (2007)	To investigate the causal factors behind injury motorcycle accidents	1,790 injury accident reports involving motorcyclists (UK, 1997–2002)	In-depth investigation	Descriptive statistics	Motorcycle DRL Reflective clothing – both	Perception problem mainly at T-junctions Such accidents seem to involve older drivers with relatively high levels of driving experience
Haque et al. (2012)	To study the effects of traffic, environmental, roadway factors on motorcycle crashes	21,922 crashes involving motorcycles (Singapore, 2004–08)	Observational study	Statistical analysis (Log-linear model)		Night-time influence increases crash risk particularly during merging and diverging manoeuvres on expressways, and turning manoeuvres at intersections

Table 1.1   Continued

Author	Objective	Data	Method	Methodology	Conspicuity configuration	Findings
Jenness et al. (2011)	To determine the impact of car DRL use on motorcycle crashes	Canada and northern US crash databases	Observational study	Statistical analysis (Logistic regression)	DRL use for other vehicles	Widespread use of DRL in the vehicle fleet increases the relative risk for certain types of multi-vehicle motorcycle crashes.
Pai (2009)	To examine how motorists' failure to give way affects motorcyclist injury severity at priority T-junctions	34,783 CMV injury accidents at T-junctions UK 1991–2004	Observational study	Statistical modelling (binary logit)	Motorcyclists' attributes Crash characteristics Lighting conditions Motorcycle size	Injuries are greater when travelling-straight motorcycle on the main road crashes into a right-turn car from the minor road Teenaged motorists predispose riders to a greater injury risk in angle perpendicular crashes Injuries to riders are greatest in angle oblique collisions with elderly motorists
Pai et al. (2009)	To examine the characteristics of automobile-motorcycle gap-acceptance accidents at priority T-junctions	38,096 CMV injury accidents at T-junctions (UK, 1991–2005)	Observational study	Statistical analysis (Mixed logit)	Motorcyclists' attributes Crash characteristics Lighting conditions Motorcycle size	Motorcycles' ROW is more likely to be violated on non-built-up roads and in diminished light conditions Elderly and female motorists appear to be over-represented in gap-acceptance crashes

Table 1.1 Continued

Author	Objective	Data	Method	Methodology	Conspicuity configuration	Findings
Preusser et al. (1995)	To analyse fatal motorcycle crashes by crash type	N=2704 fatal crashes occurring in the USA during 1992	Observational study	Classification and descriptive statistics		Motorcycle accidents involving an interaction with other vehicles account for over 26% of total crashes.
						They occur more often at intersections, in urban areas, during times of the day when more traffic would be expected
						Typically the motorcyclist has the ROW and some other vehicle fails to grant this ROW moving into the path of the motorcycle
Saleh et al. (2010)	To evaluate the interaction between PTW accidents and infrastructure	National databases (2005–07) from GR, IT, UK, ES, DE, AT. In-depth data from ES, DE, AT	Observational study + In-depth study	Macroscopic and Microscopic accident analyses; In-depth studies with road infrastructure data	Driving environment. Road infrastructure relevant contributing factors. Light condition/visib-ility	Specific outputs on critical intersection types, light conditions, infrastructural/design elements
						Visibility/conspicuity regarding road infrastructure
						Cross-European investigation on PTW crash causes
						Spanish data identify roundabouts in night-time as critical regarding conspicuity
						Motorcyclists often experience reduced visibility when wearing glasses, visors or wind shields

Table 1.1 Continued

Author	Objective	Data	Method	Methodology	Conspicuity configuration	Findings
Shaheed et al. (2011)	To investigate the effect of potential motorcycle conspicuity related factors on motorcycle crash severity outcomes	3,693 car-motorcycle crashes (Iowa, 2001–08)	Observational study	Contingency tables Multinomial logit	Lighting conditions Rural/urban environment	Severe injury outcomes in: Dark conditions Angle crashes with the car turning left and the motorcycle moving straight Dry surface conditions
Umar et al. (1996)	To explore the contributory factors of conspicuity-related motorcycle accidents	4,958 conspicuity-related motorcycle accidents (Malaysia, 1991–93)	Observational study	Statistical analysis (GLM)	DRL for motorcycles	Running headlight reduces conspicuity-related motorcycle accidents by about 29%
Wells et al. (2004)	To investigate whether motorcycle crash injuries are associated with conspicuity	463 motorcycle riders (cases) involved in injury accidents 1,233 riders (controls)	Population-based case-control study	Statistical analysis	Reflective or fluorescent clothing; headlight operation; colour of helmet; colour of clothing from waist up; colour of clothing from waist down; colour of motorcycle	Wearing reflective or fluorescent clothing and white or light coloured helmets and using headlights in daytime could reduce injury crashes by up to one-third

that lead to behavioural changes. On the contrary, population-based case-control studies are very well suited but remain few in number. In-depth databases are rare but deliver a clearer picture – more research and investigations are needed. A limited number of in-depth investigations try to conclude on the conspicuity issues, due to specific information on the crash location combined with the crash time. Overall, *the conspicuity hypothesis testing remains inconclusive* as long as several data and methodological limitations hold. In particular, crash databases should be enriched to include conspicuity elements and more in-depth and case-control studies are needed. Finally, more targeted laboratory studies would help to understand how conspicuity treatments may improve detection and response behaviours and to minimize risks.

References

Berg, W., Berglund, E., Strang, A. and Baum, M. (2007). Attention capturing properties of high frequency luminance flicker: Implications for brake light conspicuity. *Transportation Research Part F*, 10, 22–32.

ACEM (2004). MAIDS: In-depth investigation of accidents involving powered two wheelers. Report of the Association of European Motorcycle Manufacturers, Brussels. Retrieved from: http://www.maids-study.eu

Alferdinck, J. (2004). Evaluation of emergency lights display. Technical report of TNO Human Factors Research Institute, Report No TM-04-C020.

Brooks, A., Chiang, D., Smith, T., Zellner, J., Peters, J., Compagne, J. (2005). A driving simulator methodology for evaluating enhanced motorcycle conspicuity. In Proceedings of the 19th International Technical Conference on the Enhanced Safety of Vehicles, Paper No. 05–0041-O, Washington, DC.

Burg, A. and Beers, J. (1978). Reflectorization for night-time conspicuity of bicycles and motorcycles. *Journal of Safety Research*, 10, 69–77.

Cavallo, V., Pinto, M. (2012). Are car running lights detrimental to motorcycle conspicuity? *Accident Analysis & Prevention*, 49, 78–85.

Cercarelli, L., Arnold, P., Rosman, D., Sleet, D. and Thornett, M. (1992). Travel exposure and choice of comparison crashes for examining motorcycle conspicuity by analysis of crash data. *Accident Analysis & Prevention*, 24, 363–8.

Clabaux, N., Brenac, T., Perrin, C., Magnin, J., Canu, B. and Van Elslande, P. (2012). Motorcyclists' speed and 'looked-but-failed-to-see' accidents. *Accident Analysis & Prevention*, 49, 73–7.

Clarke, D., Ward, P., Bartle, C. and Truman, W. (2007). The role of motorcyclist and other driver behaviour in two types of serious accidents in the UK. *Accident Analysis & Prevention*, 39, 974–81.

Crundall, D. (2011). Visual attention while driving: Measures of eye movements used in driving research. In Porter, E.B. (ed.) *Handbook of Traffic Psychology*, 137–48. London, UK: Academic Press.

Crundall, D., Bibby, P., Clarke, D. and Bartle, C. (2008). Car drivers' attitudes towards motorcyclists: A survey. *Accident Analysis & Prevention*, 40, 983–93.

Crundall, D., Crundall, E., Clarke, D. and Shahar, A. (2012). Why do car drivers fail to give way to motorcycles at T-junctions? *Accident Analysis & Prevention*, 44, 88–96.

De Lapparent, M. (2006). Empirical Bayesian analysis of accident severity for motorcyclists in large French urban areas. *Accident Analysis & Prevention*, 38 (2), 260–268.

Farmer, C. and Williams, A. (2002). Effects of daytime running lights on multiple-vehicle daylight crashes in the United States. *Accident Analysis & Prevention*, 34, 197–203.

Fulton, E., Kirby, C. and Stroud, P. (1980). Daytime motorcycle conspicuity. Transport and Road Research Laboratory, Supplementary report 625, USA.

Gershon, P., Ben-asher, N. and Shinar, D. (2012). Attention and search conspicuity of motorcycles as a function of their visual context. *Accident Analysis & Prevention*, 44, 97–103.

Gershon, P. and Shinar, D. (2013). Increasing motorcycle attention and search conspicuity by using Alternative-Blinking Lights System. *Accident Analysis & Prevention*, 50, 801–10.

Gould, M., Poulter, D., Helman, S. and Wann, J. (2012a). Errors in judging the approach rate of motorcycles in night-time conditions and the effect of an improved lighting configuration. *Accident Analysis & Prevention*, 45, 432–7.

Gould, M., Poulter, D., Helman, S. and Wann, J. (2012b). Judgments of approach speed for motorcycles across different lighting levels and the effect of an improved tri-headlight configuration. *Accident Analysis & Prevention*, 48, 341–5.

Haque M.M., Chin H.C. and Huang, H. (2009). Modeling fault among motorcyclists involved in crashes. *Accident Analysis & Prevention*, 41, 327–35.

Haque, M., Chin, H. and Debnath, A. (2012). An investigation on multi-vehicle motorcycle crashes using log-linear models. *Safety Science*, 50, 352–62.

Haworth, N. and Debnath, A.K. (2013). How similar are two-unit bicycle and motorcycle crashes? *Accident Analysis & Prevention*, 58, 15–25.

Helman, S., Weare, A., Palmer, M. and Fernandez-Medina, K. (2012). Literature review of interventions to improve the conspicuity of motorcyclists and help avoid 'looked but failed to see' accidents. Transportation Research Laboratory, Project report No. PPR638 New Zealand.

Herslund, M. and Jorgensen, N. (2003). Looked-but-failed-to-see-errors in traffic. *Accident Analysis & Prevention*, 35, 885–91.

Hole, G., Tyrrell, L. and Langham, M. (1996). Some factors affecting motorcyclists' conspicuity. *Ergonomics*, 39, 946–65.

Horswill, M., Helman, S., Ardiles, P. and Wann, J. (2005). Motorcycle accident risk could be inflated by a time to arrival illusion. *Optometry and Vision Science*, 82, 740–746.

Huang, B. and Preston, J. (2004). A literature review on motorcycle collisions. Oxford University, UK: Transport Studies Unit.

IRTAD (2009). Road safety annual report. Retrieved from: http://www.irfnet.ch/files-upload/knowledges/IRTAD-ANNUAL-REPORT_2009.pdf

IRTAD (2013). Road safety annual report. Retrieved from: http://www.internationaltransportforum.org/irtadpublic/pdf/13IrtadReport.pdf

Jenness, J., Jenkins, F. and Zador, P. (2011). Motorcycle conspicuity and the effect of fleet DRL: Analysis of two-vehicle fatal crashes in Canada and the United States 2001–2007. Report No. DOT HS 811 505, National Highway Traffic Safety Administration.

Kim, K. and Boski, J. (2001). Finding fault in motorcycle accidents in Hawaii: Environmental, temporal, spatial, and human factors. *Transportation Research Record*, 1779, 182–188.

Kraus, J.F. (1989). Epidemiology of head injury. In Cooper, P.R. (ed.), *Head Injury*, 1–19. Baltimore, MD: Williams & Wilkins.

Lacherez, P., Wood, J., Marszalek, R. and King, M. (2013). Visibility-related characteristics of crashes involving bicyclists and motor-vehicles – responses from an online questionnaire study. Transportation Research Part F, 20, 52–8.

Lalani, N. and Holden, E. (1978). The Greater London, 'Ride Bright' campaign – its effect on motorcyclist conspicuity and casualties. *Traffic Engineering and Control*, 404–7.

Lenné, M. and Mitsopoulos-Rubens, E. (2011). Driver's decision to turn across the path of a motorcycle with low beam headlights. In Proceedings of the Human Factors and Ergonomics Society Annual Meeting, 55, 1850–1853.

Lin, M.-R. and Kraus, J. (2009). A review of risk factors and patterns of motorcycle injuries. *Accident Analysis & Prevention*, 41, 710–722.

Madsen, J., Andersen, T. and Lahrmann, H. (2013). Safety effects of permanent running lights for bicycles: A controlled experiment. *Accident Analysis & Prevention*, 50, 820–829.

Magazzù, D., Comelli, M. and Marinoni, A. (2006). Are car drivers holding a motorcycle license less responsible for motorcycle-car crash occurrence? A non-parametric approach. *Accident Analysis & Prevention*, 38, 365–70.

Mitsopoulos-Rubens, E. and Lenné, M. (2012). Issues in motorcycle sensory and cognitive conspicuity: the impact of motorcycle low-beam headlights and riding experience on drivers' decisions to turn across the path of a motorcycle. *Accident Analysis & Prevention*, 49, 86–95.

Muller, A. (1982). An evaluation of the effectiveness of motorcycle daytime headlight laws. *American Journal of Public Health*, 72, 1136–41.

NHTSA (2007). *Traffic Safety Facts 2005: Motorcycles*. National Highway Traffic Safety Administration, Washington, DC.

NPRA – Statens vegvesen (Norwegian Public Roads Administration), (2004). Handbook 245e: MC Safety – Design and Operation of Roads and Traffic Systems.

Olson, P., Halstead-Nussloch, R. and Sivak, M. (1981). The effect of improvements in motorcycle/motorcyclist conspicuity on driver behaviour. *Human Factors*, 23, 237–48.

Pai, C.-W. (2009). Motorcyclist injury severity in angle crashes at T-junctions: Identifying significant factors and analysing what made motorists fail to yield to motorcycles. *Safety Science*, 47, 1097–160.

Pai, C.-W. (2011). Motorcycle right-of-way accidents – A literature review. *Accident Analysis & Prevention*, 43, 971–82.

Pai, C.-W., Huang, K. and Saleh, W. (2009). A mixed-logit analysis of motorists' right-of-way violation in motorcycle accidents at priority T-junctions. *Accident Analysis & Prevention*, 41, 565–73.

Pai, C.-W. and Saleh, W. (2007). An analysis of motorcyclist injury severity under various traffic control measures at three-legged junctions. *Safety Science*, 45, 832–47.

Pai, C.-W. and Saleh, W. (2008). Exploring motorcyclist injury severity in approach-turn collisions at T-junctions: Focusing on the effects of driver's failure to yield and junction control measures. *Accident Analysis & Prevention*, 40, 479–86.

Perlot, A. and Prower, S. (2003). Review of the evidence for motorcycle and motorcar daytime lights. Retrieved from: http://www.lightsout.org/docs/perlot -prower-DL3331a.pdf

Preusser, D., Williams, A. and Ulmer, R. (1995). Analysis of fatal motorcycle crashes: crash typing. *Accident Analysis & Prevention*, 27, 845–51.

Ragot-Court, I., Munduteguy, C. and Fournier, J.-Y. (2012). Risk and threat factors in prior representations of driving situations among powered two-wheeler riders and car drivers. *Accident Analysis & Prevention*, 49, 96–104.

Ramsey, J. and Brinkley, W. (1977). Enhanced motorcycle noticeability through daytime use of visual signal warning devices. *Journal of Safety Research*, 9, 77–84.

Rogé, J., Douissembekov, E. and Vienne, F. (2012). Low conspicuity of motorcycles for car drivers: dominant role of bottom-up control of visual attention or deficit of top-down control? *Human Factors*, 1, 14–25.

Rößger, L., Hagen, K., Krzywinski, J. and Schlag, B. (2012). Recognisability of different configurations of front lights on motorcycles. *Accident Analysis & Prevention*, 44, 82–7.

Saleh, P., Golias J., Yannis G. et al. (2010). FP7 project 2-BE-SAFE, Deliverable 1.2, Interaction between Powered Two-Wheeler Accidents and Infrastructure. Retrieved from: http://www.2besafe.eu/sites/default/files/deliverables/2BES_ D28_GuidelinesPolicyRecommendationsAndFurtherResearchPriorities.pdf

Rumar, K. (1980). Running lights, conspicuity, glare and accident reduction. *Accident Analysis & Prevention*, 12, 151–7.

Savolainen, P. and Mannering, F. (2007). Probabilistic models of motorcyclists' injury severities in single- and multi-vehicle crashes. *Accident Analysis and Prevention*, 39, 955–63.

Shaheed, M., Zhang, W., Gkritza, K. and Hans, Z. (2011). Differences in motorcycle conspicuity related factors and motorcycle crash severities in daylight and dark conditions. In Proceedings of the 3rd International Conference on Road Safety and Simulation, Indianapolis, USA.

Shaheed, M., Gkritza, K. and Marshall, D. (2012). Motorcycle conspicuity – what factors have the greatest impacts. Center for Transportation Research and Education, Iowa State University, MTC Project 2011–01.

Smither, J. and Torrez, L. (2010). Motorcycle conspicuity: Effects of age and daytime running lights. *Human Factors*, 52, 355–69.

Thomson, G.A. (1980). The role frontal conspicuity has in road accidents. *Accident Analysis & Prevention*, 12, 165–78.

Tyrrell, R., Wood, J. and Carberry, T. (2004). On-road measures of pedestrians' estimates of their own nighttime conspicuity. *Journal of Safety Research*, 35, 483–90.

Umar, R., Murray, G. and Hills, B. (1996). Modelling of conspicuity-related motorcycle accidents in Serembn and Shah Alam, Malaysia. *Accident Analysis & Prevention*, 28, 325–32.

Van Elslande, P., Yannis, G., Feypell, V., Papadimitriou, E., Tan, C. and Jordan, M. (2013). Contributory factors of powered two wheelers crashes. In Proceedings of the 13th WCTR, Rio de Janeiro, Brazil.

Vlahogianni, E., Yannis, G. and Golias, J. (2012). Overview of critical risk factors in Power-Two-Wheeler safety. *Accident Analysis & Prevention*, 49, 12–22.

Walton, D. (2010). Car vs. motorcycle accidents: A critical examination of the literature. Opus central laboratories, Report No. 528062.00.

Walton, D., Buchanan, J. and Murray, S. (2013). Exploring factors distinguishing car-versus-car from car-versus-motorcycle in intersection crashes. Transportation Research Part F, 17, 145–53.

Watts, G. (1980). The evaluation of conspicuity aids for cyclists and motorcyclists. In Osborne, D.J. and Levis, J.A. (eds) *Human Factors in Transport Research*. London, UK: Academic Press.

Wells, S., Mullin, B., Norton, R., Langley, J., Connor, J., Lay-Yee, R. and Jackson, R. (2004). Motorcycle rider conspicuity and crash related injury: Case-control study. *British Medical Journal*, 328, 857–60.

Williams, M. and Hoffmann, E. (1977). *The Influence of Motorcycle Visibility on Traffic Accidents*. University of Melbourne, Australia: Department of Engineering.

Williams, M. and Hoffmann, E. (1979). Motorcycle conspicuity and traffic accidents. *Accident Analysis & Prevention*, 11, 209–24.

Wood, J., Tyrrell, R., Marszalek, R. and Lacherez, P. (2013). Bicyclists overestimate their own night-time conspicuity and underestimate the benefits of retroflective markers on the moveable joints. *Accident Analysis & Prevention*, 55, 48–53.

Wulf, G., Hancock, P. and Rahimi, M. (1989). Motorcycle conspicuity: An evaluation and synthesis of influential factors. *Journal of Safety Research*, 20, 153–76.

Yuan, W. (2000). The effectiveness of the 'ride-bright' legislation for motorcycles in Singapore. *Accident Analysis & Prevention*, 32, 559–63.

Zador, P. (1985). Motorcycle headlight-use laws and fatal motorcycle crashes in the US, 1975–83. *American Journal of Public Health*, 75, 543–6.

Chapter 2

Psychological Factors in Seeing Motorcycles

Vanessa Beanland, Michael G. Lenné and Lars Rößger

One of the leading causes of motorcycle crashes are right-of-way violations in which another road user, typically a car driver, turns across the path of an oncoming motorcycle (Clarke et al., 2007). Before discussing psychological theories that could explain the human factors underlying this crash type, it is instructive to consider a basic behavioural taxonomy and failure classification of the typical errors involved in car-motorcycle crashes. Crundall, Humphrey and Clarke (2008) proposed three key behaviours that drivers must execute in order to avoid collisions: first, they must look at the motorcycle; second, they must detect its presence; and finally, they must appraise it appropriately. *Detection errors* arise from failure of one of the first two stages, and comprise the majority of crashes and involve other road users failing to detect the motorcycle or detecting it too late to avoid a collision (Pai, 2011). Such crashes are referred to as 'look but fail to see' crashes because afterwards drivers often indicate that they looked at the rider's location but failed to detect the motorcycle. In contrast, *decision errors* arise from inappropriate appraisal; these errors occur when another road user detects a motorcycle but misjudges its speed and/or location and consequently accepts a gap that is too small (Pai, 2011). Both detection and decision errors frequently occur in conditions of good visibility, when drivers presumably have the best possible chance to detect and appraise an oncoming vehicle. Aggregated crash data indicates that the 'modal' motorcycle crash occurs during the day in urban areas, good weather conditions and involves an experienced driver failing to see a motorcycle and consequently turning across the rider's path (Hancock et al., 2005). This suggests drivers' difficulties in perceiving and accurately appraising motorcycles result at least partially from cognitive factors; if they were purely perceptual limitations then they would be most common under conditions of poor visibility.

Parallels can be drawn between the detection and decision errors that occur in car-motorcycle crashes and several psychological phenomena that have been observed in controlled laboratory experiments. The aim of this chapter is to review psychological literature relating to detection and decision errors, respectively, and to highlight how these theoretical accounts contribute to our understanding of why these crashes occur and what can be done to address them. The main focus will be on 'look but fail to see' errors; there are several psychological phenomena that involve this type of error, which are collectively referred to as induced failures of visual awareness (Simons and Rensink, 2003). The final section will focus on time-to-arrival judgements, which form a crucial aspect of decision errors, and factors that can lead observers to make erroneous time-to-arrival judgements. Basic psychological research provides an understanding of the mechanisms underlying both 'look but fail to see' and time-to-arrival judgement errors, which can in turn help us to develop appropriate countermeasures and strategies for preventing or minimising the severity of the types of crashes that typically result from detection or decision errors.

Detection Errors: 'Look But Fail To See' Motorcycle Crashes

There are many psychological phenomena that give rise to 'look but fail to see' errors. The term 'look but fail to see' is a broad designation commonly used by safety practitioners and researchers to describe a crash situation in which the driver fails to detect a clearly visible object or hazard. Although it is often referred to as being a specific (singular) type of error, there are in fact many psychological phenomena that could result in a driver apparently looking without 'seeing' what is before their eyes. Before considering the role of these psychological factors, it is helpful to first briefly review the nature of the problem as observed in motorcycle crashes. Researchers have been discussing the problem of 'look but fail to see' motorcycle crashes since at least the 1960s (Olson et al., 1981; Williams and Hoffman, 1979b). 'Look but fail to see' crashes are not exclusive to motorcycles; other research has indicated that drivers may 'look but fail to see' even highly salient vehicles such as marked police cars (Langham et al., 2002) and passenger trains (Salmon et al., 2013). However, 'look but fail to see' errors are most commonly discussed in relation to two-wheeled vehicles because they are implicated as a leading cause of motorcycle crashes (ACEM, 2009), and crash analyses suggest that 'look but fail to see' errors are more common with motorcycles compared to other vehicles (Brooks et al., 2005; Van Elslande et al., 2012).

The exact prevalence of 'look but fail to see' crashes is difficult to estimate since this information is typically only obtained in relatively small in-depth crash investigations, and even then it is difficult to ascertain with certainty. In some samples, 50 per cent or more of at-fault car drivers report that they collided with a motorcycle because they did not see it (Brooks et al., 2005; Clabaux et al., 2012; Van Elslande et al., 2012; Williams and Hoffman, 1979b). There is some evidence that self-reports may over-represent 'look but fail to see' crashes, perhaps because 'looking without seeing' implies less negligence than either failing to look or seeing the motorcycle but misjudging the situation (Crundall et al., 2012). In one study 64 per cent of drivers claimed that they 'look but fail to see' a motorcycle, but after ruling out other contributing factors, including visual obstructions occluding the motorcycle, the researchers estimated that only 21 per cent of crashes involved a driver failing to see a motorcycle that was in clear view (Williams and Hoffman, 1979b).

Given the consistent reports of 'look but fail to see' crashes, researchers have attempted to reduce these errors by devising treatments that improve the physical salience or conspicuity of motorcycles and motorcycle riders. Conspicuity enhancements can include daytime running lights (DRLs), uniquely coloured or shaped lighting configurations, and reflective, fluorescent or brightly coloured clothing or fairings. Research on the effectiveness on these conspicuity enhancements has consistently demonstrated that headlight treatments such as daytime running lights can improve drivers' detection of motorcycles (for example Olson et al., 1981; Smither and Torrez, 2010; Thomson, 1980; Williams and Hoffman, 1979a), although this effect is reduced when other vehicles also use daytime running lights (Cavallo and Pinto, 2012). Other conspicuity treatments tend to yield small or inconsistent effects on motorcycle detection. Experimental evidence suggests that the effectiveness of 'high-visibility' rider clothing and fairings depends on their contrast with the background (Gershon et al., 2012; Hole et al., 1996; Williams and Hoffman, 1979a). As such, the lack of effect observed in real-world settings is likely to due to the fact that contrast is variable and changes depending on the environment, lighting and weather conditions.

There is little real-world data evaluating the effectiveness of conspicuity enhancements in preventing 'look but fail to see' crashes. One study of UK police records reported that at least 30 per cent of riders struck in right-of-way violations (which often, but do not always, result from 'look but fail to see' errors) were using some form of conspicuity enhancement (Clarke et al., 2007). Unfortunately this study did not provide data on the use of conspicuity enhancements more generally, and did not distinguish between different types of conspicuity enhancement (for example, daytime running lights versus reflective clothing) so it is not possible to assess how effective these treatments were. A case-control study of motorcycle crashes in New Zealand found that both wearing white versus black helmets and wearing reflective or fluorescent clothing were associated with reduced crash risk, whereas the colour of the motorcycle itself and wearing light versus dark clothes had no effect (Wells et al., 2004). This suggests that some (though not all) conspicuity enhancements may improve other road users' detection of motorcycles, but since the authors did not explicitly examine 'look but fail to see' errors there is no conclusive evidence that conspicuity enhancements specifically reduce 'look but fail to see' crashes. An alternative could be that conspicuity enhancements are associated with lower overall crash risk because they are adopted by riders who engage in safer riding behaviour. Regardless, there is persuasive evidence that even if existing conspicuity enhancements do reduce 'look but fail to see' crashes, they do not eliminate them. Similarly, 'look but fail to see' crashes most frequently occur under conditions that are arguably the optimal conditions for detecting motorcycles: during daytime, fine weather and relatively uncluttered rural environments (Clabaux et al., 2012; Hancock et al., 2005). 'Look but fail to see' errors are the result of cognitive limitations, rather than being purely physical perceptual errors, and for this reason the following sections examine the psychological literature in this domain and highlight the potential contributions of this literature of our understanding of 'look but fail to see' crash causation and countermeasures.

Failure of Visual Awareness

There are several types of induced failures of visual awareness, in which observers are unable to detect a stimulus that would otherwise be clearly visible because of the specific task demands imposed on them. These psychological phenomena have been referred to as the 'dark side of visual attention' because they focus on circumstances in which objects fail to capture the observer's attention (Chun and Marois, 2002), in contrast to the vast majority of visual attention research, which focuses on the 'bright side' of how and why objects capture attention (this research is discussed in Chapter 4). Specific examples include: *inattentional blindness*, the failure to detect unexpected objects (Mack and Rock, 1998; Simons and Chabris, 1999); *change blindness*, the failure to detect expected or unexpected changes (Rensink et al., 1997); *attentional blink*, the failure to detect the second of two targets in a rapid sequence (Raymond et al., Arnell, 1992); and *repetition blindness*, the failure to detect the second instance of a repeated target (Kanwisher, 1987).

Different types of induced blindness are often discussed in conjunction with each other (for example, Chun and Marois, 2002; Enns and Di Lollo, 2000; Kim and Blake, 2005; Lavie, 2006; Rensink, 2000; Wolfe, 1999) and some subtypes appear to be closely related. For example, individuals who are more susceptible to attentional blink are also more likely to experience inattentional blindness (Beanland and Pammer, 2012). There are also numerous similarities between change blindness and inattentional blindness, particularly for change blindness paradigms that involve detection of unexpected changes (Jensen et al., 2011). However, there are also important conceptual distinctions between these phenomena. These distinctions are highly relevant to researchers investigating failures of visual awareness in laboratory settings, since they influence methodological choices, but they are also relevant to anyone seeking to transfer findings from experimental research to real-world problems. Several theoretical explanations have been put forth to explain failures of visual awareness, but most account for a single phenomenon; for example, processing 'bottleneck' models of attentional blink (Dux and Marois, 2009). Ultimately, the diversity of paradigms and theories suggests that there is no single mechanism by which observers fail to detect a target; 'look but fail to see' errors are not a unitary phenomenon, but could arise from several distinct cognitive limitations.

Among the various phenomena that involve failures of visual awareness, inattentional blindness and change blindness have the most direct relevance to motorcycle crashes. Both are relatively easy to induce in laboratory experiments using a variety of stimuli, ranging from simple shapes and letters (Koivisto and Revonsuo, 2007, 2009; Mack and Rock, 1998; Pashler, 1988; Scholl, 2000) to more complex stimuli including photographs, simulations and film clips (Haines,

1991; Pammer and Blink, 2013; Rensink et al., 1997; Shinoda et al., 2001; Simons and Chabris, 1999). More importantly, both inattentional blindness and change blindness have been demonstrated to occur in real-world interactions (Chabris et al., 2011; Hyman et al., 2010; Simons and Levin, 1998). In contrast, attentional blink and repetition blindness occur reliably in computerised tasks but have less obvious real-world correlates, because they result from limitations in the temporal distribution of attention and occur only when two targets are presented within 800 msec of each other. As such, the following sections will focus on inattentional blindness and change blindness.

Inattentional Blindness

Inattentional blindness is defined as failure to detect an unexpected object or event when one's attention is directed elsewhere (Mack and Rock, 1998). In a typical inattentional blindness experiment participants will be given an attention-consuming primary task, such as simultaneously tracking multiple moving objects. The primary task is usually designed to engage visual attention (for example, Mack and Rock, Most et al., 2001) but can engage other processes including working memory (Fougnie and Marois, 2007), auditory attention (Beanland et al., 2011; Pizzighello and Bressan, 2008), speech generation (Scholl et al., 2003) and physical exertion (Hüttermann and Memmert, 2012). After several trials of the primary task by itself, participants then complete a critical trial in which an additional, unexpected stimulus appears. Individuals who cannot report the presence of this stimulus are classified as experiencing inattentional blindness. The final trial of the experiment is the full attention trial, for which the participant is instructed to simply watch the display without completing the initial primary task. The purpose of the full attention trial is to demonstrate that the unexpected stimulus was easily visible and can be detected in the absence of the attention-consuming primary task.

The unexpected stimuli in an inattentional blindness experiment may be either static or dynamic. In static paradigms the unexpected stimulus is added to the visual display for the entire trial, but trials are usually very brief; typically 200 msec or less (Mack and Rock, 1998). The advantage of such brief presentations is that they allow a high degree of control over stimuli; the brevity of stimulus presentation means that observers do not have an opportunity to execute eye movements. However, such displays lack ecological validity. For this reason, sustained inattentional blindness paradigms were developed to investigate situations in which observers fail to detect a dynamic unexpected stimulus that appears for several seconds. Sustained inattentional blindness paradigms use both computerised tasks, such as object-tracking (Beanland and Pammer, 2010, 2012; Beanland et al., 2011; Koivisto and Revonsuo, 2008; Most et al., 2005; Most et al., 2000; Most et al., 2001; Simons and Jensen, 2009) and naturalistic videos of scenarios such as unexpected interlopers in basketball and handball games (Furley et al., 2010; Memmert and Furley, 2007; Neisser, 1979; Neisser and Becklen, 1975; Simons, 2010; Simons and Chabris, 1999), or a woman entering the room and scraping her nails down a chalkboard (Wayand et al., 2005). Some studies have also examined inattentional blindness in real-world interactions, by staging events such as a unicycling clown in a public square (Hyman et al., 2010) or a man being beaten (Chabris et al., 2011) and then investigating what factors increase or decrease the likelihood of individuals noticing these events.

Despite the methodological differences between static and dynamic inattentional blindness paradigms, they are considered equivalent phenomena and produce similar findings. Static and dynamic inattentional blindness paradigms differ mainly in terms of their attentional demands; this has implications for theoretical explanations of inattentional blindness. Static paradigms demand focused attention to complete the primary task, but not necessarily selective attention, since the observer expects that the entire brief display will be task-relevant. In contrast, dynamic displays typically include distractor items, meaning that the participant must selectively attend to some items and ignore others. Distractors play an important role in inattentional blindness: some observers fail to detect unexpected stimuli even without distractors but the rate of inattentional blindness is significantly higher when distractors are present (Koivisto and Revonsuo, 2008). The number of task-irrelevant items has no effect – one distractor or five will yield similar rates of inattentional blindness – the crucial factor is that observers have something to ignore (Koivisto and Revonsuo, 2008). Driving is therefore more comparable to dynamic inattentional blindness tasks, since it involves selective attention to a moving visual environment; most if not all of the time when we are driving, there will be more information in the environment than we can absorb. Often a lot of information will be irrelevant, requiring drivers to selectively ignore the irrelevant visual information (for example, billboards advertising services) and selectively focus on the most relevant information (for example, other vehicles and potential hazards).

Several findings from inattentional blindness research have particular relevance to motorcycle crashes. Eye-tracking studies have revealed that eye movements do not differ between noticers and non-noticers of the unexpected stimulus (Beanland and Pammer, 2010; Koivisto et al., 2004; Memmert, 2006; Richards et al., 2011). Observers can fixate directly on the unexpected stimulus for up to 1 sec and still be unable to report its presence (Memmert, 2006), even if the unexpected stimulus is highly conspicuous, such as a moving red cross in an otherwise monochromatic display (Richards et al., 2011). These studies confirm that 'looking without seeing' is a real, relatively common phenomenon: many observers fail to notice what is literally right before their eyes. Furthermore, they reveal a double dissociation between looking and seeing (Beanland and Pammer, 2010). It is possible to fixate one's gaze on an object without noticing it, but it is also possible to notice an object without fixating it by covertly deploying attention to one's peripheral vision. This has implications for how people scan the environment while driving; simply glancing at your blind spot is no guarantee that you will notice what is there. During their learner phase novice drivers are often reminded to physically turn and look at various areas, but simply looking is not sufficient unless the driver is also deliberately attending to that region and prepared for the possibility of unexpected hazards appearing.

Most 'look but fail to see' motorcycle crashes involve right-of-way violations in which the driver turns across the path of an oncoming motorcycle; since the motorcycle would have been located directly ahead of the driver, it is a reasonable assumption that it appeared centrally within the driver's field of view. Similarly, high rates of inattentional blindness can occur for unexpected stimuli that appear centrally within the observer's 'attentional zone', including for objects that overlap or directly interact with task-relevant objects that the observer is tracking (Most et al., 2000; Simons and Chabris, 1999). Some studies have found that inattentional blindness is greatest either at or immediately surrounding fixation (Mack and Rock, 1998; Thakral and Slotnick, 2010; but see Koivisto et al., 2004), suggesting that observers suppress parafoveal input when they are attending to their fixation point but suppress foveal information when they are attending elsewhere. In laboratories studies the location of the unexpected stimulus predicts its likelihood of detection: rates of noticing decrease systematically as the stimulus increases in distance from the observer's attentional focus (Most et al., 2000; Newby and Rock, 1998). Equivalent effects have not been observed in crash situations, but that is more likely due to the nature of the traffic flow; if a motorcycle is not located near the driver then they are unlikely to come into conflict, so even if drivers fail to detect the motorcycle there will be no adverse consequences.

Aside from location, several stimulus characteristics influence the likelihood of inattentional blindness. The absolute physical properties of an unexpected stimulus do not greatly increase or decrease its likelihood of detection; when viewing a greyscale display observers are equally likely to detect a white stimulus, a black stimulus or a bright-red one (Most et al., 2001). However, the relative properties of an unexpected stimulus are extremely important. Observers are less likely to detect objects that share similar features with objects that they are ignoring (Most et al., 2001, 2005); a simulator study found that drivers were more likely to collide with a motorcycle if it was the same colour as navigational symbols that they were instructed to ignore (Most and Astur, 2007). Conversely, observers are more likely to notice objects that are semantically or physically similar to whatever they are paying attention to (Most et al., 2001) or objects that have personal relevance (Mack and Rock, 1998).

The inattentional blindness literature has a number of important implications for motorcycle safety. First, it confirms that 'look but fail to see' errors are genuine and in fact can be extremely common, as evidenced by the various eye movement studies revealing that inattentional blindness can occur even when the observer directly fixates the unexpected stimulus. Second, it indicates that physical salience is not sufficient for an unexpected object to capture attention, which suggests that simply increasing a motorcycle's sensory conspicuity will not eliminate driver's difficulties in detecting riders. The findings regarding the role of semantic similarity and personal relevance in inattentional blindness are consistent with applied research indicating that 'dual drivers' who also ride motorcycles are less likely to be at-fault in car-motorcycle crashes (Magazzù et al., 2006) and exhibit more cautious behaviour around motorcycles (Crundall et al., 2012; Shahar et al., 2012; but see Ohlhauser et al., 2011). It is worth noting that the advantage conferred on dual licence holders appears to be highly specific: motorcycle riders are better at detecting other motorcycles and motorcycle-relevant hazards, even while driving a car, but do not have superior hazard perception abilities overall (Crundall et al., 2013).

Change Blindness

Change blindness describes a failure or delay in detecting a change to a visual stimulus or scene, which occurs when the changes take place in a moment during brief global disruptions of the retinal image (for example, McConkie and Currie, 1996; Pashler, 1988; Rensink et al., 1997; Simons and Levin, 1998). It is assumed that the local transient produced by the change, which would normally attract an observer's attention, is masked due to this retinal disruption. Interruptions of retinal information occur naturally due to eye movements and blinks, resulting in *gaze-contingent* change blindness (Grimes, 1996; O'Regan et al., 2000). Change blindness occurs during eye movements due to *saccadic suppression*, which means that the visual resolution during saccades is drastically reduced: information capturing is suppressed during the 30–40 msec preceding and up to 120 msec after the initiation of a saccade (Volkmann et al., 1997). Change blindness may also occur during eye blinks, when the eye is briefly physically occluded by the eyelid. The average blink rate for healthy human beings is between 12–15 blinks/min but can vary between 2–25 blinks/min (Barbato et al., 2000). The occurrence of change blindness is not restricted to gaze-contingent disruptions; it can also be induced by brief artificial disruptions to the visual scene. Artificial disruptions of the retinal image may be induced by 'flickers' between the stimulus and a blank screen (for example, Rensink et al., 1997), physical occlusion of the visual scene by a moving object (Simons and Levin, 1998), scene cuts in films (Simons and Levin, 1997) or even simultaneously presented distractors that do not occlude the change, such as 'mudsplashes' on a car window (Bahrami, 2003; O'Regan et al., 1999; see Velichkovsky et al., 2002, for a comparison of change blindness methods).

The visual change of interest in a change blindness study may be expected or unexpected (in contrast to inattentional blindness, where by definition the critical stimulus of interest is always unexpected), but a crucial aspect of change blindness is that once the observer realises the change they consider it surprising that they ever failed to detect it (Jensen et al., 2011). In both inattentional blindness and change blindness, once an observer identifies the stimulus of interest it becomes obvious and they cannot return to a state of blindness, meaning that the same stimulus cannot be used repeatedly. Change blindness occurs for a wide range of visual stimuli including simple arrays of digits (Pashler, 1988), photographs (Rensink et al., 1997) and, most strikingly, a person in a real-life conversation (Simons and Levin, 1998).

As with inattentional blindness, change blindness depends on the absence of attention: observers are faster to detect changes to items of central interest in their visual scene or current task, compared to items of marginal interest (Rensink et al., 1997). For example, drivers are less likely to notice changes to road signs when they are completing a car-following task, compared to 'normal' driving in which they must attend to road signs in order to obey road rules (Shinoda et al., 2001). Conversely, observers are better at detecting changes to items that were previously cued (Rensink et al., 1997) or items that captured exogenous attention, such as late-onset stimuli (Scholl, 2000). Semantic content and task-relevance also influence the likelihood and speed of detecting changes, because observers are more likely to attend to items that have greater personal or task-relevance and this attention reduces change blindness. For example, insomniacs are more likely to detect changes to sleep-related stimuli (Marchetti et al., 2006) and substance users are more likely to detect changes to drug or alcohol-related items (Jones et al., 2003), both of which have strong personal relevance to the groups in question. Several studies of change blindness while driving have also found that drivers are more likely to detect changes to task-relevant stimuli (Galpin et al., 2009; Shinoda et al., 2001; Velichkovsky et al., 2002). In particular, drivers are faster at detecting changes to task-relevant objects, such as road signs, compared to task-irrelevant changes such as to nearby buildings or walls (Galpin et al., 2009). When changes are made to road signs, drivers are also more likely to detect changes to road signs that appear at intersections, compared to mid-block (Shinoda et al., 2001). These results are further supported by findings from Velichkovsky et al. (2002), who also found faster and more accurate change detection for task-relevant stimuli in road scenes, although they noted that change blindness was stronger in dynamic stimuli compared to static stimuli, especially when change blindness was induced by saccadic suppression.

Despite the similarities between inattentional blindness and change blindness they are distinct phenomena with some notable differences (Rensink, 2000). Task demands differ substantially between inattentional and change blindness paradigms. Inattentional blindness occurs when the observer is engaged in another task, but when observers do not have to perform this task they can perceive the unexpected stimulus without difficulty (Mack and Rock, 1998). In change blindness paradigms, detecting the change often *is* the observer's primary task (that is, if the change is expected) and yet they still experience 'blindness'. Consequently, change blindness can occur even when the change is expected (Rensink et al., 1997). Whereas the experience of inattentional blindness occurs independently of eye movements and thus represents a genuine case of 'looking without seeing', instances of gaze-contingent change blindness actually rely on eye movements and as such change blindness may result from the failure to look rather than 'look but fail to see' (or failing to look again, which would be necessary in order to detect that a stimulus has changed). Finally, although both inattentional and change blindness result from inattention or insufficient attention to a stimulus, change blindness involves a stronger memory component because it involves the failure to notice that an object has changed over time, whereas inattentional blindness involves completely failing to detect the presence of the object in the first place (Rensink, 2000).

So what does this mean for motorcycle safety? Failure to detect an oncoming motorcycle could be construed as either inattentional blindness or change blindness, depending on the precise sequence of events. We could imagine a driver approaching an intersection, slowing to a stop and then searching the road ahead to determine whether any vehicles were approaching. If, under these circumstances, the driver failed to detect an oncoming motorcycle directly in his field of view, this would constitute an example of inattentional blindness. Alternatively, our driver might approach an unsignalzed T-intersection intending to turn from a minor road into a major road. In this case, the driver might make an initial check to the right, correctly deduce the way is clear, then check to the left, see that is clear, then check back to the right and fail to notice a recently appeared oncoming motorcycle. This would be an instance of change blindness; the driver fails to detect that the view to the right has changed since he initially checked that direction, and there is now a motorcycle approaching. Overall, it is plausible that both inattentional and change blindness could contribute to real-world instances of drivers failing to detect motorcycles and in practice it would be challenging to conclusively determine which error contributed to a specific crash. This is particularly true of crash investigations that rely on *post-hoc* self-report, since drivers may remember or be aware of the sequence of events preceding the crash. It might be possible to infer the occurrence of inattentional blindness or change blindness during naturalistic driving studies or on-road studies using instrumented vehicles (although the latter, being of shorter duration, would be less likely to record a crash or near-crash event), since these methods involve recording continuous video footage of both the driver and the surrounding environment.

Theoretical Explanations for Looking Without Seeing

The preceding section reviewed empirical literature regarding failures of visual awareness, particularly inattentional blindness and change blindness. Several theories have been developed to explain these phenomena, as well as other instances of observers failing to detect fully visible stimuli. Understanding these theories is instructive because they provide specific predictions as to the conditions that are most or least likely to induce 'look but fail to see' errors. The following subsections review the theories that are most relevant to inattentional blindness in particular, since it is the archetypal example of 'look but fail to see' errors in experimental psychology.

The Perceptual Cycle and Schema Theory

One of the first modern theoretical attempts to explain 'looking without seeing' was Neisser's (1976) perceptual cycle, which has formed the basis of much subsequent work in both cognitive psychology (Most et al., 2005) and ergonomics (Plant and Stanton, 2013). Although most research concentrates on visual perception and the visual content of our schemata, Neisser (1976) intended for the perceptual cycle model to explain perceptual experience across all sensory processes. Traditional information processing models of cognition equate the mind with a computer: visual images received by the retina are processed in a series of stages until it reaches consciousness. These models fail to explain why two people looking at an identical scene will experience a different conscious percept, or why we pick up some information but not others; Neisser's perceptual cycle model was an attempt to explain these anomalies.

According to the perceptual cycle model, perception is a constructive process that begins with anticipatory schemata (see Figure 2.1). Schemata direct our active exploration of the environment; we focus our attention depending on what we expect to encounter, which is based on our past experiences and knowledge

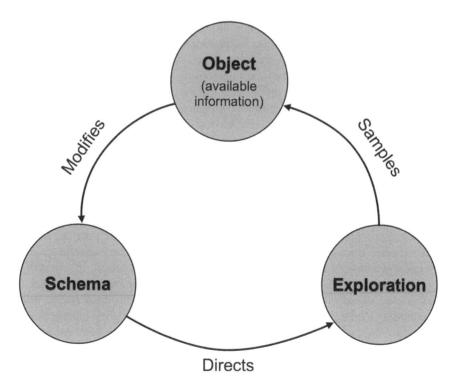

Figure 2.1   Neisser's (1976) perceptual cycle

(Neisser, 1976). Different levels of schemata exist, so depending on which schema is active a driver may be prepared to see 'something', a vehicle, a motorcycle or a motorcycle approaching in the bus lane. Schemata contain both physical and semantic features, so when we see a motorcycle we do not merely perceive a small metal object with a human on top; we perceive a road user that we must be mindful of in case our paths conflict. It is important to note that schemata do not directly determine perception; rather, they determine our exploration of the environment, and our perception is therefore the product of our schema interacting with the information available from the environment. This model therefore explains why, for example, observers' patterns of eye movements differ depending on what they are searching for, even within identical visual scenes (see Chapter 3): in the real world we are frequently confronted by 'rich' perceptual environments that support the engagement of multiple schemata. Indeed, Neisser argues that schemata will be most influential in such situations, where the scene is ambiguous or there are multiple interpretations possible: the schema that we adopt determines how we perceive the scene.

Our perception of the world is a constructive, ongoing process. We use anticipatory schemata to direct our attention (that is, where we will explore perceptually). This in turn determines which objects we perceive, and the information that we pick up from those objects will be used to modify the original schema so that the cycle can continue in future.

As a result of our schema-directed exploration of the environment, we sample or pick up information about particular objects, which we then use to modify schemata for further exploration. Thus perception is not merely constructive but also iterative, with increasing information gained over time. This explains why skilled and unskilled observers differ in both their eye movements (for example, Humphrey and Underwood, 2009; Underwood, 2007) and the amount and nature of information they perceive; for example, basketball experts are less likely than other athletes to experience inattentional blindness for unexpected events during a basketball game (Memmert, 2006). Skilled observers have more experience sampling the environment, which can help them to develop more detailed schemata. Neisser (1976) argues that although there is a limit to the amount of information we can perceive and process, this limit is not fixed and can be increased through training. Each new perceptual experience is also an opportunity to develop a new schema, through perceptual learning. What we expect (schema) guides what we perceive, but we do not only perceive what is expected; if something violates our schema and we notice the violation, that will prompt us to engage in further exploration of the perceptual environment in order to integrate the new information. Consequently the schema that we develop for a given situation will depend on what we perceive in that situation, which in turn will be based on our initial anticipatory schema.

The perceptual cycle model provides a straightforward explanation of 'look but fail to see' errors: observers fail to notice unexpected objects because they are inconsistent with their schema, and observers may also fail to detect other objects or changes because their schema is inadequate or inappropriate for the situation. Notably Neisser (1976, 1979) argued that there is no special mechanism or filter that screens out unattended information; rather than choosing what to ignore, Neisser believed that perception is an active process and we only perceive the information that we actively select (as determined by our schema). This can be easily understood using apple picking as an analogy: 'To pick one apple from a tree you need not filter out all the others; you just don't pick them. A theory of apple picking would have much to explain … but it would not have to specify a mechanism to keep unwanted apples out of your mouth' (Neisser, 1976: 85). In other words, according to Neisser's perceptual cycle the reason why drivers sometimes fail to detect motorcycles is that motorcycles are not appropriately represented in their schemata, meaning that they focus their search on looking for the most commonly expected features or objects (for example, cars). There is empirical support for this notion from a number of observational and on-road studies. For example, when stopped at intersections drivers scan regions of space where other cars might appear, but often neglect areas that are unlikely to contain cars but may contain other road users, such as bicycles or pedestrians (Summala et al., 1996).

The concept of schema is not unique to the perceptual cycle model; schemata have been applied to describe and explain a range of phenomena including memory (Pezdek et al., 1989) and cognitive biases observed in psychiatric disorders (Young et al., 2003). Schema theories have strong explanatory power but have been subject to some criticisms (see Plant and Stanton, 2013, for a review). These criticisms primarily relate to the fact that, being cognitive constructs, schemata cannot be directly observed. Studies may be particularly vulnerable to this criticism if they use schema as a *post-hoc* explanation; that is, if an observer misses some information and it is inferred that this was due to the activation of an inadequate or inappropriate schema (Plant and Stanton, 2013). However, it is possible to more systematically measure schema content by using verbal protocols, in which the participant 'thinks aloud' while completing a specific task, such as driving (Walker et al., 2011). Verbal protocol analysis reveals the individual's situation awareness, which includes both the information that they perceive and the information that they seek (that is, schema).

The strength of the perceptual cycle model is that it explains why drivers fail to detect motorcycles, since it focuses on how inappropriate schemata are developed and maintained.

Several studies suggest that schema content varies between different road user groups, such that individuals focus on the information that they perceive to be most relevant, depending on their mode of transport. For example, car drivers focus on other cars, motorcyclists focus more on hazard anticipation, cyclists focus more on navigating safely through congested traffic situations (Salmon et al., 2013; Walker et al., 2011). This suggests an obvious solution: to reduce motorcycle crashes, we need to modify drivers' schemata and make them more acutely aware of motorcycles. The flaw with the perceptual cycle model is that it does not provide concrete guidance on how this can be achieved; conversely, the self-reinforcing nature of schemata actually suggests that even if drivers started out with an appropriate schema, over time the schema would shift due to infrequent interaction with motorcycles. Although it seems counterintuitive, empirical evidence confirms that increased experience can have a detrimental effect. Homogenous or highly specific experiences lead the individual to develop rigid expectancies, which results in impaired performance in related but noticeable different situations (Van Elslande and Faucher-Alberton, 1997).

Attentional Set

Although Neisser's (1976) perceptual cycle provides a solid foundation for understanding failures of visual awareness, it is not a complete explanation. For instance, Neisser explained that an unexpected stimulus can initiate the perceptual cycle (that is, at the object phase, thus generating a new schema for future exploration) but did not specify how or why an unexpected stimulus breaks into the cycle. In order to bridge this explanatory gap, Most et al. (2005) reformulated the perceptual cycle to accommodate the concept of attentional set. *Attentional set* refers to the idea that observers can 'set' their attention according to the features or information that they expect will be presented, which forms a top-down constraint on attention (Folk et al., 1992). These top-down settings can override the bottom-up salience of visual objects, as demonstrated in visual search tasks that contain an additional singleton; that is, a non-target that is featurally distinct from both the target and distractors. Because of their salience, singletons capture attention and slow search times by diverting attention away from the target (Theeuwes, 1992) unless the observer can adopt an accurate attentional set for differentiating the target, such as knowing which colour to look for (Folk et al., 1992).

Expectations play a powerful role in determining both attentional set and inattentional blindness. Research on inattentional blindness indicates that observers set their attention based on what they expect to encounter in the primary task. If these expectations are met, then they are more likely to experience inattentional blindness (White and Aimola Davies, 2008), which suggests that observers terminate their perceptual exploration of the environment when their expectations are met. However, if their expectations about the primary task are violated in some way (for example, they are shown more or fewer items than they expected; White and Aimola Davies, 2008), then they are more likely to detect unexpected objects, because the violation of expectations prompts further exploration of the visual environment.

Attentional set explains the findings that unexpected objects are significantly less likely to be noticed when they are featurally distinct from attended objects, even if the unexpected object is highly salient such as a bright-red cross on a grey background (Koivisto and Revonsuo, 2008; Most et al., 2001, 2005). Detection rates drop when the unexpected object is similar to distractor items, even reaching zero if the unexpected stimulus and distractors are identical (Most et al., 2001). This is because observers have set their attention to ignore items with a specific feature (for example, 'ignore everything black'). Similarly, the finding that sustained inattentional blindness can occur in the absence of distractors (Koivisto and Revonsuo, 2008) makes more sense in the context of attentional set: participants set their attention to focus on expected items and fail to notice an unexpected object simply because it is not part of their attentional set. Collectively these findings suggest two key functions of attentional set: to focus on attended objects and to screen out irrelevant objects. Observers can set their attention according to the expected number (White and Aimola Davies, 2008), semantic category (Koivisto and Revonsuo, 2007, 2009), luminance or shape of stimuli (Most et al., 2001, 2005), as well as for more complex features such as racial identity of faces (Most et al., 2005). In the context of driving, it is likely that most drivers set their attention to detect cars (because cars comprise the most common and obvious hazards on the road), which makes them more efficient at detecting cars and car-like vehicles, but impairs their ability to detect hazards that are dissimilar to cars – such as motorcycles.

The literature concerning attentional set implies that the best way to avoid 'look but fail to see' errors would be to adopt a broad attentional set; to 'expect the unexpected' at all times. However, this strategy may not be entirely practical, for two reasons. First, many tasks explicitly demand focused attention and require that observers adopt a specific attentional set. In most urban driving environments we are surrounded by irrelevant information; if we tried to absorb all of it then it may increase our likelihood of detecting unexpected objects, but our driving performance would be impaired. Second, recent evidence suggests that individuals adopt an attentional set based not on what they are told to look for, but rather according to what they actually experience in the real world. For example, if observers are required to track a set of black square objects, they could be instructed to set their attention according to either colour (black), shape (square) or both (black square). Theoretically, if they were to set their attention based on shape then they should be more likely to notice a white square, compared to a black cross. In actuality, observers are unlikely to notice the white square because despite the task instructions, they adopt an attentional set for colour rather than shape; it is easier to differentiate black versus white than square versus circle (Aimola Davies et al., 2013). Thus even if drivers are reminded to search for non-car vehicles, they may inadvertently fall back into adopting an attentional set that only includes cars, because cars are so ubiquitous and easier to find.

Perceptual Load

Another theory that has been applied to explain failures of visual awareness is Lavie's (1995) *perceptual load theory*, which was proposed as a compromise between traditional early selection (for example, Broadbent, 1958) and late selection (for example, Deutsch and Deutsch, 1963) views of attention. Load theory argues that processing capacity is limited, consistent with early selection theories of attention, but also that processing is automatic, consistent with late selection views. Therefore, task-relevant stimuli are processed first, but processing of irrelevant stimuli continues until capacity is exhausted. The implication is that when the perceptual load of task-relevant items is low, unexpected stimuli are more likely to be detected because the observer has sufficient 'leftover' capacity to process irrelevant stimuli. Increases in load can involve perceptual load (that is, more items in the display, as in Lavie's original work) or cognitive load (that is, the same display with a more demanding task). When load is increased more resources are required for the primary task, meaning fewer resources are available for processing task-irrelevant or unexpected items.

Several studies have demonstrated that more demanding primary tasks are associated with decreased detection of unexpected items (Cartwright-Finch and Lavie, 2007). Increasing visual short term memory load (Todd et al., 2005) and increasing the number of task-relevant items in the display both increase inattentional blindness (Koivisto and Revonsuo, 2008; White and Aimola Davies, 2008). Increasing the speed or difficulty of the primary task also increases inattentional blindness (Beanland and Pammer, 2010; Neisser, 1979; Simons and Chabris, 1999; Simons and Jensen, 2009). Conversely, practicing the primary task decreases the likelihood of experiencing inattentional blindness (Neisser, 1979; Richards et al., 2010). The effects of load are not restricted to visual load; increasing executive working memory load also increases inattentional blindness (Fougnie and Marois, 2007) and some research suggests that cross-modal sources of perceptual load, such as talking on a mobile phone, can also induce inattentional blindness (Pizzighello and Bressan, 2008; Scholl et al., 2003). Interestingly, the semantic congruency effect, whereby observers are more likely to detect an unexpected stimulus belonging to the same category as task-relevant items, remains significant even when primary task demands increase substantially (Koivisto and Revonsuo, 2009). This suggests that unexpected objects are still submitted for semantic processing under conditions of high load, even though they do not reach awareness.

Load theory does a reasonable job of predicting general circumstances under which inattentional blindness will occur – namely, when task load is high. In addition, task demands have been demonstrated to modulate the role of attentional set and expectations. Observers' expectations have little effect on inattentional blindness when perceptual load is low, but when load is high expectations significantly determine whether an observer will detect an unexpected stimulus (White and Aimola Davies, 2008). However, load theory does not actually explain inattentional blindness. In particular, that meaningful stimuli appear impervious to inattentional blindness is beyond the scope of load theory. This is because load theory applies only to the primary task demands, rather than addressing characteristics of the unexpected stimulus itself. Furthermore, some recent research has suggested that increasing task load to a moderate level can actually reduce inattentional blindness (Beanland et al., 2011; Hüttermann and Memmert, 2012), implying that there is U-shaped relationship between load and inattentional blindness. When load is very high, inattentional blindness is more likely because observers do not have sufficient cognitive resources to detect unexpected stimuli. However, when load is too low then observers may disengage and thus be more likely to experience inattentional blindness, even though they have ample capacity to detect unexpected items. Overall, these findings suggest that there are a number of conditions when drivers are most likely to experience 'look but fail to see' errors: when first learning to drive; when driving under difficult conditions, such as in the rain; when talking on a mobile phone; but also when driving in relatively easy and familiar conditions, when it is possible to 'switch off'.

Search Efficiency

The theories discussed above focus on explaining 'look but fail to see' errors in experimental paradigms that are primarily designed to examine failures of visual awareness, particularly inattentional blindness. Another avenue for explaining failures of awareness is through theories of visual search, although these theories focus more on how observers successfully detect targets rather than why they miss them (as is the case with inattentional blindness). Visual search paradigms require an observer to confirm the presence or absence of a predetermined target in a visual array or scene. The primary dependent variable in most visual search experiments is reaction time, or how long it takes observers to decide whether a target is present or absent, and classic theories of visual search are based almost exclusively on reaction time data (for example, Treisman and Gelade, 1980). Reaction time is used rather than accuracy because in most laboratory experiments accuracy is extremely high. Recently it has been revealed that high search accuracy may be an artefact of experimental design. In traditional visual search paradigms the target appears on 50 per cent of trials, so the observer has reasonable certainty that the target will be present and consequently they have more incentive to search until they find it (Wolfe et al., 2005). Altering target prevalence so that it appears more or less frequently affects accuracy and, most notably, when target prevalence is very low (1–2 per cent of trials) observers are more likely to miss targets; this finding has been dubbed the low prevalence effect (Rich et al., 2008; Van Wert et al., 2009; Wolfe et al., 2005, 2007).

The low prevalence effect has been explained in terms of a *quitting threshold* (Wolfe, 2007). Observers do not search exhaustively until they either find the target or have scrutinised the entire display. Rather, it appears that they spend a reasonable amount of time searching for each target, but this time varies depending on the nature of the task and their recent search history. For example, when observers are aware that they have made a mistake they slow their response time (spend longer searching) on subsequent trials, presumably to avoid making further mistakes (Chun and Wolfe, 1996). When targets are low prevalence, compared to high prevalence, observers respond faster on 'target absent' trials but are also more likely to miss targets (Wolfe et al., 2007). This is because the target is usually absent, so they adopt a lower quitting threshold. A higher quitting threshold (that is, longer duration spent searching for the target) would be inefficient since it would consume a lot more time and in most cases the resulting decision (that is, target absent) would be the same. The low prevalence effect has been observed for a range of stimuli, ranging from simplistic arrays of letters or shapes (Rich et al., 2008) to airport luggage (Wolfe et al., 2005, 2007; Van Wert et al., 2009) and cytology screening (Evans et al., 2011).

It has been suggested that the low prevalence effect could explain (or partly explain) drivers' difficulties in detecting motorcycles; this possibility is explored further in Chapter 10. Most research examining the conspicuity of motorcycles and other powered two-wheelers has been conducted in developed countries where powered two-wheelers constitute a 'minority vehicle' (Haworth, 2012). Based on existing evidence, it seems highly plausible that the low prevalence of powered two-wheelers could exacerbate drivers' difficulties in detecting these vehicles in countries such as the UK, Australia and the USA, where motorcycles comprise approximately 1 per cent of traffic. Low prevalence is clearly not the only cause of 'look but fail to see' motorcycle crashes; however, in future research it would be worthwhile exploring whether the nature and extent of 'look but fail to see' errors differs in countries such as Malaysia, where motorcycles are considerably more common.

Decision Errors: Seeing Without Comprehending

The preceding discussion focuses on circumstances that give rise to detection errors, which contribute to a large proportion of car-motorcycle crashes, but as noted this is not the only type of error that drivers may experience (Pai, 2011). Aside from completely failing to see the motorcycle (as in 'look but fail to see' errors), drivers may see the motorcycle but inappropriately appraises its features (Crundall et al., 2008). The most problematic errors occur when a driver misperceives the motorcycle's time-to-arrival and turns across its path when the motorcycle is too close. One study notes that a frequent comment drivers make following right-of-way violation crashes is, 'I thought I could make it. But the motorcycle came unexpectedly fast' (Nagayama et al., 1980: 958). Although much discussion focuses on vehicle speed, appraisal errors can arise from either underestimating the speed of an oncoming motorcycle, overestimating its distance or a combination of the two.

Several experimental designs have been used to assess drivers' appraisal errors. These include speed judgements, time-to-arrival judgements and gap-acceptance tasks. In *speed judgement* tasks, the observer must either estimate the speed of an oncoming vehicle or compare its speed to that of a reference vehicle (Gould et al., 2012a, 2012b). This type of task can provide insights into how vehicle characteristics influence speed judgements. In time-to-arrival judgements, the observer views a dynamic display of an approaching object or vehicle, which is occluded (typically for 1–7 seconds) before it reaches the observer. The observer is required to press a button to indicate when the occluded object would reach them (Horswill et al., 2005). Since objects appear larger as they move closer, observers estimate time-to-arrival based on *tau*, the rate of optical expansion of an object (Lee et al., 2009). Finally, in *gap-acceptance tasks* observers are required to drive through a simulated intersection, or view a road scene, in which they have to decide whether it is safe to turn or cross a road during a given traffic gap (Crundall et al., 2008; Nagayama et al., 1980). Gap-acceptance tasks have ecological validity but are less useful for judging perceptual biases; if a driver accepts a shorter gap in front of a motorcycle compared to a car, this could be that she perceived the motorcycle as farther away or it could be that she correctly judged the motorcycle's distance but simply decided it was appropriate to accept the shorter gap for some other reason (Nagayama et al., 1980).

Most research on speed and time-to-arrival judgements indicates that, compared to cars and other four-wheeled vehicles, drivers perceive motorcycles as being slower and farther away, with a longer time-to-arrival (Gould et al., 2012a, 2012b; Horswill et al., 2005). Speed judgement for motorcycles is particularly impaired at night-time; one study found that observers judged a reference car travelling at 30 mph to be equivalent in speed to a motorcycle travelling at approximately 150 mph (Gould et al., 2012a). There is a linear association between vehicle size and time-to-arrival, whereby smaller vehicles are consistently judged to arrive later than larger vehicles (Horswill et al., 2005). Briefly occluding the oncoming vehicle can reduce time-to-arrival estimates (DeLucia, 2004; Horswill et al., 2005), which would presumably lead to more conservative gap-acceptance decisions, but this depends on the duration and nature of the occluder. Longer occlusion durations (4–7 seconds) are associated with reduced time-to-arrival estimates, but shorter occlusions of 1–2 seconds are associated with increased time-to-arrival estimates for motorcycles (Horswill et al., 2005). This is concerning given that most traffic situations are likely to involve relatively brief occluders. If the occluding object is moving, such as a car passing in front of an oncoming motorcycle, then depending on the direction of its movement it may either increase or decrease the observer's estimated time-to-arrival for the motorcycle (DeLucia, 2004). Overall these results suggest that time-to-arrival illusions and speed misperceptions are likely to contribute to right-of-way violations against motorcycles.

Beyond the misjudgement of the time-to-arrival for motorcycles compared to other vehicles, Oberfeld and Hecht (2008) recently raised a further question which seems highly relevant for drivers' estimations towards oncoming motorcycles in real-world scenarios. The authors questioned whether the visual system is capable of isolating information that is relevant for time-to-arrival judgements of a target object in the context of multiple distractors or whether the time-to-arrival information of other objects would distort the judgement of the relevant target. In other words: how does another slower vehicle on a parallel trajectory to the motorcycle affect observer's time-to-arrival judgement for the motorcycle, and beyond this, how does it affect the observer's gap-acceptance decision regarding the motorcycle? An initial hypothesis was that the time-to-arrival information for target and distractor might be averaged which, in turn, would imply an increased safety-critical behaviour towards the motorcycle in our example. Their results confirm a response bias resulting from the presence of a late-arriving distractor, but the effect was in the opposite direction to what was predicted: later arriving distractors caused an early response towards the target. This suggests that in situations where motorcycles are surrounded by slower-moving traffic, drivers will be even more likely to overestimate the motorcyclist's time-to-arrival and consequently more likely to turn across its path. Oberfeld and Hecht assume that the task-related significance of the distractor might determine how time-to-arrival-information for multiple objects is processed. Further research might explore the role of task-related significance in more detail and the implication for traffic safety.

Whereas detection errors (especially 'look but fail to see' errors) involve failure of both top-down and bottom-up attention processes, there is strong evidence that decision errors result largely from bottom-up processes. When drivers are allowed unlimited time to view a road scene, their gap-acceptance decisions are equivalent regardless of whether the oncoming vehicle is a car or a motorcycle (Crundall et al., 2008). Novel headlight configurations that emphasise the motorcycle's horizontal width also significantly improve observers' estimates of motorcycle speeds, especially at night-time (Gould et al., 2012a, 2012b). Similar headlight configurations have been found to improve the detectability of motorcycles (Rößger et al., 2012), although given the robustness of inattentional blindness and related phenomena it is possible that headlight treatments alone may be insufficient to completely eliminate 'look but fail to see' errors.

Conclusion

Psychological literature has much to offer us in terms of understanding car-motorcycle crashes. Broadly, the errors that drivers make in relation to motorcycles can be categorised as either detection errors (that is, 'look but fail to see') or decision errors (that is, gap-acceptance errors). Inattentional blindness research, and to a lesser extent change blindness research, provides useful findings that can be extrapolated to drivers' detection errors regarding motorcycles. Although there is no one theory that comprehensively explains why observers sometimes fail to detect fully visible stimuli, existing theories do provide valuable insight into why these errors occur and how they could be minimized. Importantly, all of the research indicates that simply increasing the sensory conspicuity of motorcycles (for example, by changing headlight configurations or adding high-visibility treatments) may reduce conspicuity-related crashes but will probably not eliminate them. This is because the low conspicuity of motorcycles is not solely due to their small size; it is exacerbated by the fact that they are typically rare and unexpected.

Attentional set and the perceptual cycle theory both suggest a role for driver training that fosters development of more complex and diverse driving-related schemata. On the other hand, perceptual load theory and its relationship to inattentional blindness highlight the need to consider the context in which individuals drive and the need to prevent drivers from engaging in both external distractions (such as talking on a mobile phone) and internal distraction (such as drifting off into task-unrelated thoughts). Both driver training and distraction are effectively cognitive interventions for the driver; they aim to change the way the driver approaches the driving task. In contrast, the research related to decision errors suggests that observers have strong biases that affect their ability to judge the speed and time-to-arrival of motorcycles. Given that these biases arise from physical properties of visual objects, the most effective methods for eliminating them appear to be through changing the physical appearance of the motorcycle. Existing research indicates that this will reduce, but not eliminate, drivers' tendencies to overestimate the time-to-arrival of oncoming motorcycles. Technological interventions such as collision warning systems could also be beneficial in reducing both detection and decision errors, although given the robustness of inattentional blindness effects there remains the possibility that in some circumstances drivers could still fail to detect motorcycles (that is, if they failed to detect both the motorcycle itself and the system warning). Overall, given the strength and the ubiquity of these errors, the most appropriate strategy would be to adopt multiple interventions aimed at reducing both detection and decision errors, thus minimizing their likelihood as much as possible.

References

ACEM (Association des Constructeurs Européens de Motocycles). (2009). In-depth investigations of accidents involving powered two-wheelers (MAIDS). Retrieved from: http://www.maids-study.eu/pdf/MAIDS2.pdf

Aimola Davies, A.M., Waterman, S., White, R.C. and Davies, M. (2013). When you fail to see what you were told to look for: Inattentional blindness and task instructions. *Consciousness & Cognition*, 22(1), 221–30. doi: 10.1016/j.concog.2012.11.015

Bahrami, B. (2003). Object property encoding and change blindness in multiple object tracking. *Visual Cognition*, 10(8), 949–63. doi: 10.1080/135062803440 00158

Barbato, G., Ficca, G., Muscettola, G., Fichele, M., Beatrice, M. and Rinaldi, F. (2000). Diurnal variation in spontaneous eye-blink rate. *Psychiatry Research*, 93(2), 145–51. doi: 10.1016/S0165–1781(00)00108–6

Beanland, V., Allen, R.A. and Pammer, K. (2011). Attending to music decreases inattentional blindness. *Consciousness & Cognition*, 20(4), 1282–92. doi: 10.1016/j.concog.2011.04.009

Beanland, V. and Pammer, K. (2010). Looking without seeing or seeing without looking? Eye movements in sustained inattentional blindness. *Vision Research*, 50, 977–88. doi: 10.1016/j.visres.2010.02.024

Beanland, V. and Pammer, K. (2012). Minds on the blink: The relationship between inattentional blindness and attentional blink. *Attention, Perception, & Psychophysics*, 74(2), 322–30. doi: 10.3758/s13414–011–0241–4

Broadbent, D.E. (1958). *Perception and Communication*. London: Pergamon Press.

Brooks, A.M., Chiang, D.P., Smith, T.A., Zellner, J.W., Peters, J.P. and Compagne, J. (2005). *A Driving Simulator Methodology for Evaluating Enhanced Motorcycle Conspicuity*. Paper presented at the 19th International Technical Conference on the Enhanced Safety of Vehicles, Washington, DC, 6–9 June 2005.

Cartwright-Finch, U. and Lavie, N. (2007). The role of perceptual load in inattentional blindness. *Cognition*, 102, 321–40. doi: 10.1016/j.cognition. 2006.01.002

Cavallo, V. and Pinto, M. (2012). Are car daytime running lights detrimental to motorcycle conspicuity? *Accident Analysis & Prevention*, 49, 78–85. doi: 10.1016/j.aap.2011.09.013

Chabris, C.F., Weinberger, A., Fontaine, M. and Simons, D.J. (2011). You do not talk about Fight Club if you do not notice Fight Club: Inattentional blindness for a simulated real-world assault. *i-Perception*, 2(2), 150–53. doi: 10.1068/ i0436

Chun, M. and Marois, R. (2002). The dark side of visual attention. *Current Opinion in Neurobiology*, 12, 184–9. doi: 10.1016/S0959–4388(02)00309–4

Chun, M.M. and Wolfe, J.M. (1996). Just say no: How are visual searches terminated when there is no target present? *Cognitive Psychology*, 30(1), 39–78. doi: 10.1006/cogp.1996.0002

Clabaux, N., Brenac, T., Perrin, C., Magnin, J., Canu, B. and Van Elslande, P. (2012). Motorcyclists' speed and 'looked-but-failed-to-see' accidents. *Accident Analysis & Prevention*, 49, 73–7. doi: 10.1016/j.aap.2011.07.013

Clarke, D.D., Ward, P., Bartle, C. and Truman, W. (2007). The role of motorcyclist and other driver behaviour in two types of serious accident in the UK. *Accident Analysis & Prevention*, 39, 974–981. doi: 10.1016/j.aap.2007.01.002

Crundall, D., Crundall, E., Clarke, D. and Shahar, A. (2012). Why do car drivers fail to give way to motorcycles at t-junctions? *Accident Analysis & Prevention*, 44, 88–96. doi: 10.1016/j.aap.2010.10.017

Crundall, D., Humphrey, K. and Clarke, D. (2008). Perception and appraisal of approaching motorcycles at junctions. *Transportation Research Part F*, 11, 159–67. doi: 10.1016/j.trf.2007.09.003

Crundall, D., van Loon, E., Stedmon, A. W. and Crundall, E. (2013). Motorcycling experience and hazard perception. *Accident Analysis & Prevention*, 50, 456–64. doi: 10.1016/j.aap.2012.05.021

DeLucia, P.R. (2004). Time-to-contact judgements of an approaching object that is partially concealed by an occluder. *Journal of Experimental Psychology: Human Perception & Performance*, 30(2), 287–304.

Deutsch, J.A. and Deutsch, D. (1963). Attention: Some theoretical considerations. *Psychological Review*, 70, 80–90. doi: 10.1037/h0039515

Dux, P.E. and Marois, R. (2009). The attentional blink: A review of data and theory. *Attention, Perception, & Psychophysics*, 71, 1683–1700. doi: 10.3758/APP.71.8.1683

Enns, J.T. and Di Lollo, V. (2000). What's new in visual masking? *Trends in Cognitive Sciences*, 4, 345–52. doi: 10.1016/S1364–6613(00)01520–5

Evans, K.K., Tambouret, R.H., Evered, A., Wilbur, D.C. and Wolfe, J.M. (2011). Prevalence of abnormalities influences cytologists' error rates in screening for cervical cancer. *Archives of Pathology & Laboratory Medicine*, 135(12), 1557–60. doi: 10.5858/arpa.2010–0739-OA

Folk, C.L., Remington, R.W. and Johnston, J.C. (1992). Involuntary covert orienting is contingent on attentional control settings. *Journal of Experimental Psychology: Human Perception & Performance*, 18, 1030–44. doi: 10.1037/0096–1523.18.4.1030

Fougnie, D. and Marois, R. (2007). Executive working memory load induces inattentional blindness. *Psychonomic Bulletin & Review*, 14, 142–7.

Furley, P., Memmert, D. and Heller, C. (2010). The dark side of visual awareness in sport: Inattentional blindness in a real-world basketball task. *Attention, Perception & Psychophysics*, 72, 1327–37. doi: 10.3758/APP.72.5.1327

Galpin, A., Underwood, G. and Crundall, D. (2009). Change blindness in driving scenes. *Transportation Research Part F: Traffic Psychology & Behaviour*, 12(2), 179–85. doi: 10.1016/j.trf.2008.11.002

Gershon, P., Ben-Asher, N. and Shinar, D. (2012). Attention and search conspicuity of motorcycles as a function of their visual context. *Accident Analysis & Prevention*, 44, 97–103. doi: 10.1016/j.aap.2010.12.015

Grimes, J. (1996). On the failure to detect changes in scenes across saccades. In Akins, K.A. (ed.), *Perception. Vancouver studies in cognitive science*, Vol. 5, 89–110. New York, NY: Oxford University Press.

Gould, M., Poulter, D.R., Helman, S. and Wann, J.P. (2012a). Errors in judging the approach rate of motorcycles in nighttime conditions and the effect of an improved lighting configuration. *Accident Analysis & Prevention*, 45, 432–7. doi: 10.1016/j.aap.2011.08.012

Gould, M., Poulter, D.R., Helman, S. and Wann, J.P. (2012b). Judgments of approach speed for motorcycles across different lighting levels and the effect of an improved tri-headlight configuration. *Accident Analysis & Prevention*, 48, 341–5. doi: 10.1016/j.aap.2012.02.002

Haines, R.F. (1991). A breakdown in simultaneous information processing. In Obrecht, G. and Stark, L.W. (eds), *Presbyopia Research: From Molecular Biology to Visual Adaptation*, 171–5. New York: Plenum.

Hancock, P.A., Oron-Gilad, T. and Thom, D.R. (2005). Human factors issues in motorcycle collisions. In Noy, I. and Karwovski, W. (eds), *Handbook of Human Factors in Litigation*, 18:11–20. Boca Raton, FL: CRC Press.

Haworth, N. (2012). Powered two wheelers in a changing world – Challenges and opportunities. *Accident Analysis & Prevention*, 44, 12–18. doi: 10.1016/j.aap.2010.10.031

Hole, G.J., Tyrrell, L. and Langham, M. (1996). Some factors affecting motorcyclists' conspicuity. *Ergonomics*, 39(7), 946–65. doi:10.1080/00140139608964516

Horswill, M.S., Helman, S., Ardiles, P. and Wann, J.P. (2005). Motorcycle accident risk could be inflated by a time to arrival illusion. *Optometry & Vision Science*, 82(8), 740–46.

Humphrey, K. and Underwood, G. (2009). Domain knowledge moderates the influence of visual saliency in scene recognition. *British Journal of Psychology*, 100(2), 377–98. doi: 10.1348/000712608x344780

Hüttermann, S. and Memmert, D. (2012). Moderate movement, more vision: Effects of physical exercise on inattentional blindness. *Perception*, 41(8), 963–75. doi: 10.1068/p7294

Hyman, I.E., Boss, S.M., Wise, B.M., McKenzie, K.E. and Caggiano, J.M. (2010). Did you see the unicycling clown? Inattentional blindness while walking and talking on a cell phone. *Applied Cognitive Psychology*, 24, 597–607. doi: 10.1002/acp.1638

Jensen, M.S., Yao, R., Street, W.N. and Simons, D.J. (2011). Change blindness and inattentional blindness. *Wiley Interdisciplinary Reviews: Cognitive Science*, 2(5), 529–46. doi: 10.1002/wcs.130

Jones, B.T., Jones, B.C., Smith, H. and Copley, N. (2003). A flicker paradigm for inducing change blindness reveals alcohol and cannabis information processing biases in social users. *Addiction*, 98(2), 235–44. doi: 10.1046/j.13 60–0443.2003.00270.x

Kanwisher, N.G. (1987). Repetition blindness: Type recognition without token individuation. *Cognition*, 27, 117–43. doi: 10.1016/0010–0277(87)90016–3

Kim, C.-Y. and Blake, R. (2005). Psychophysical magic: Rendering the visible 'invisible'. *Trends in Cognitive Sciences*, 9, 381–8. doi: 10.1016/j.tics.2005.06.012

Koivisto, M. and Revonsuo, A. (2007). How meaning shapes seeing. *Psychological Science*, 18, 845–9. doi: 10.1111/j.1467–9280.2007.01989.x

Koivisto, M. and Revonsuo, A. (2008). The role of unattended distractors in sustained inattentional blindness. *Psychological Research*, 72, 39–48. doi: 10.1007/s00426–006–0072–4

Koivisto, M. and Revonsuo, A. (2009). The effects of perceptual load on semantic processing under inattention. *Psychonomic Bulletin & Review*, 16, 864–868. doi: 10.3758/PBR.16.5.864

Koivisto, M., Hyönä, J. and Revonsuo, A. (2004). The effects of eye movements, spatial attention, and stimulus features on inattentional blindness. *Vision Research*, 44, 3211–21. doi: 10.1016/j.visres.2004.07.026

Langham, M., Hole, G., Edwards, J. and O'Neil, C. (2002). An analysis of 'looked but failed to see' accidents involving parked police vehicles. *Ergonomics*, 45(3), 167–85. doi: 10.1080/00140130110115363

Lavie, N. (1995). Perceptual load as a necessary condition for selective attention. *Journal of Experimental Psychology: Human Perception & Performance*, 21, 451–68. doi:10.1037/0096–1523.21.3.451

Lavie, N. (2006). The role of perceptual load in visual awareness. *Brain Research*, 1080, 91–100. doi: 10.1016/j.brainres.2005.10.023

Lee, D.N., Bootsma, R.J., Frost, B.J., Land, M., Regan, D. and Gray, R. (2009). Lee's 1976 paper. *Perception*, 38(6), 837–58. doi: 10.1068/ldmk-lee

Mack, A. and Rock, I. (1998). *Inattentional Blindness*. Cambridge, MA: MIT Press.

Magazzù, D., Comelli, M. and Marinoni, A. (2006). Are card drivers holding a motorcycle licence less responsible for motorcycle – car crash occurrence? A non-parametric approach. *Accident Analysis & Prevention*, 38, 365–70. doi: 10.1016/j.aap.2005.10.007

Marchetti, L.M., Biello, S.M., Broomfield, N.M., Macmahon, K.M.A. and Espie, C.A. (2006). Who is pre-occupied with sleep? A comparison of attention bias in people with psychophysiological insomnia, delayed sleep phase syndrome and good sleepers using the induced change blindness paradigm. *Journal of Sleep Research*, 15(2), 212–21. doi: 10.1111/j.1365–2869.2006.00510.x

McConkie, G.W. and Currie, C.B. (1996). Visual stability across saccades while viewing complex pictures. *Journal of Experimental Psychology: Human Perception & Performance*, 22, 563–81.

Memmert, D. (2006). The effects of eye movements, age, and expertise on inattentional blindness. *Consciousness & Cognition*, 15, 620–27. doi: 10.1016/j.concog.2006.01.001

Memmert, D. and Furley, P. (2007). 'I spy with my little eye!': Breadth of attention, inattentional blindness, and tactical decision making in team sports. *Journal of Sport & Exercise Psychology*, 29, 365–81.

Most, S.B., Scholl, B.J., Clifford, E.R. and Simons, D.J. (2005). What you see is what you set: Sustained inattentional blindness and the capture of awareness. *Psychological Review*, 112, 217–42. doi:10.1037/0033–295X.112.1.217

Most, S.B., Simons, D.J., Scholl, B.J. and Chabris, C.F. (2000). Sustained inattentional blindness: The role of location in the detection of unexpected dynamic events. *Psyche*, 6(14).

Most, S.B., Simons, D.J., Scholl, B.J., Jimenez, R., Clifford, E. and Chabris, C.F. (2001). How not to be seen: The contribution of similarity and selective ignoring to sustained inattentional blindness. *Psychological Science*, 12, 9–17. doi: 10.1111/1467–9280.00303

Most, S. B. and Astur, R.S. (2007). Feature-based attentional set as a cause of traffic accidents. *Visual Cognition*, 15, 125–32.

Nagayama, Y., Morita, T., Miura, T., Watanabe, J. and Murakami, N. (1980). Speed judgement of oncoming motorcycles. In *Proceedings of the International*

Motorcycle Safety Conference, Volume 2, 955–71. Linthicum, MD: Motorcycle Safety Foundation.

Neisser, U. (1976). *Cognition and Reality: Principles and Implications of Cognitive Psychology*. San Francisco, CA: Freeman.

Neisser, U. (1979). The control of information pickup in selective looking. In Pick, A.D. (ed.), *Perception and its Development: A tribute to Eleanor J. Gibson*, 201–19. Hillsdale, NJ: Lawrence Erlbaum.

Neisser, U. and Becklen, R. (1975). Selective looking: Attending to visually specified events. *Cognitive Psychology*, 7, 480–494. doi: 10.1016/0010–0285 (75)90019–5

Newby, E.A. and Rock, I. (1998). Inattentional blindness as a function of proximity to the focus of attention. *Perception*, 27, 1025–40. doi: 10.1068/p271025

Oberfeld, D. and Hecht, H. (2008). Effects of a moving distractor object on time-to-contact judgments. *Journal of Experimental Psychology: Human Perception and Performance*, 34(3), 605–23. doi: 10.1037/0096–1523.34.3.605

Ohlhauser, A.D., Milloy, S. and Caird, J.K. (2011). Driver responses to motorcycle and lead vehicle braking events: The effects of motorcycling experience and novice versus experienced drivers. *Transportation Research Part F: Traffic Psychology & Behaviour*, 14, 472–83. doi: 10.1016/j.trf.2011.08.003

Olson, P.L., Halstead-Nussloch, R. and Sivak, M. (1981). The effect of improvements in motorcycle/motorcyclist conspicuity on driver behavior. *Human Factors*, 23(2), 237–48. doi: 10.1177/001872088102300211

O'Regan, J.K., Deubel, H., Clark, J.J. and Rensink, R.A. (2000). Picture changes during blinks: Looking without seeing and seeing without looking. *Visual Cognition*, 7(1–3), 191–211. doi: 10.1080/135062800394766

O'Regan, J.K., Rensink, R.A. and Clark, J.J. (1999). Change-blindness as a result of 'mudsplashes'. *Nature*, 398(6722), 34. doi: 10.1038/17953

Pai, C.-W. (2011). Motorcycle right-of-way accidents – A literature review. *Accident Analysis & Prevention*, 43, 971–82. doi: 10.1016/j.aap.2010.11.024

Pammer, K. and Blink, C. (2013). Attentional differences in driving judgments for country and city scenes: Semantic congruency in inattentional blindness. *Accident Analysis & Prevention*, 50, 955–63. doi: 10.1016/j.aap.2012.07.026

Pashler, H. (1988). Familiarity and visual change detection. *Perception & Psychophysics*, 44, 369–78.

Pezdek, K., Whetstone, T., Reynolds, K., Askari, N. and Dougherty, T. (1989). Memory for real-world scenes: The role of consistency with schema expectation. *Journal of Experimental Psychology: Learning, Memory, & Cognition*, 15(4), 587–95. doi: 10.1037/0278–7393.15.4.587

Pizzighello, S. and Bressan, P. (2008). Auditory attention causes visual inattentional blindness. *Perception*, 37, 859–66. doi: 10.1068/p5723

Plant, K.L. and Stanton, N.A. (2013). The explanatory power of Schema Theory: Theoretical foundations and future applications in Ergonomics. *Ergonomics*, 56(1), 1–15. doi: 10.1080/00140139.2012.736542

Raymond, J.E., Shapiro, K.L. and Arnell, K.M. (1992). Temporary suppression of visual processing in an RSVP task: An attentional blink? *Journal of Experimental Psychology: Human Perception & Performance*, 18, 849–60. doi: 10.1037/0096–1523.18.3.849

Rensink, R.A. (2000). When good observers go bad: Change blindness, inattentional blindness, and visual experience. *Psyche*, 6(9).

Rensink, R.A., O'Regan, K. and Clark, J.J. (1997). To see or not to see: The need for attention to perceive changes in scenes. *Psychological Science*, 8, 368–73. doi: 10.1111/j.1467–9280.1997.tb00427.x

Rich, A.N., Kunar, M.A., Van Wert, M.J., Hidalgo-Sotelo, B., Horowitz, T.S. and Wolfe, J.M. (2008). Why do we miss rare targets? Exploring the boundaries of the low prevalence effect. *Journal of Vision*, 8(15). doi: 10.1167/8.15.15

Richards, A., Hannon, E.M. and Derakshan, N. (2010). Predicting and manipulating the incidence of inattentional blindness. *Psychological Research*, 74, 513–23. doi: 10.1007/s00426–009–0273–8

Richards, A., M. Hannon, E. and Vitkovitch, M. (2012). Distracted by distractors: Eye movements in a dynamic inattentional blindness task. *Consciousness & Cognition*, 21(1), 170–76. doi: 10.1016/j.concog.2011.09.013

Rößger, L., Hagen, K., Krzywinski, J. and Schlag, B. (2012). Recognisability of different configurations of front lights on motorcycles. *Accident Analysis & Prevention*, 44, 82–7. doi: 10.1016/j.aap.2010.12.004

Salmon, P.M., Read, G., Stanton, N.A. and Lenné, M.G. (2013). The crash at Kerang: Investigating systemic and psychological factors leading to unintentional non-compliance at rail level crossings. *Accident Analysis & Prevention*, 50, 1278–88.

Salmon, P.M., Young, K.Y. and Cornelissen, M. (2013). Compatible cognition amongst road users: The compatibility of driver, motorcyclist, and cyclist situation awareness. *Safety Science*, 56, 6–17. doi: 10.1016/j.ssci.2012.02.008

Scholl, B.J. (2000). Attenuated change blindness for exogenously attended items in a flicker paradigm. *Visual Cognition*, 7, 377–96. doi: 10.1080/135062800394856

Scholl, B.J., Noles, N.S., Pasheva, V. and Sussman, R. (2003). Talking on a cellular telephone dramatically increases 'sustained inattentional blindness'. *Journal of Vision*, 3(9), 156a. doi: 10.1167/3.9.156

Shahar, A., van Loon, E., Clarke, D. and Crundall, D. (2012). Attending overtaking cars and motorcycles through the mirrors before changing lanes. *Accident Analysis & Prevention*, 44, 104–10. doi: 10.1016/j.aap.2011.01.001

Shinoda, H., Hayhoe, M.M. and Shrivastava, A. (2001). What controls attention in natural environments? *Vision Research*, 41(25–6), 3535–45. doi: 10.1016/S0042–6989(01)00199–7

Simons, D.J. (2010). Monkeying around with the gorillas in our midst: Familiarity with an inattentional-blindness task does not improve the detection of unexpected events. *i-Perception*, 1, 3–6. doi: 10.1068/i0386

Simons, D.J. and Chabris, C.F. (1999). Gorillas in our midst: Sustained inattentional blindness for dynamic events. *Perception*, 28, 1059–74. doi: 10.1068/p2952

Simons, D.J. and Jensen, M.S. (2009). The effects of individual differences and task difficulty on inattentional blindness. *Psychonomic Bulletin & Review*, 16, 398–403. doi: 10.3758/PBR.16.2.398

Simons, D.J. and Levin, D.T. (1997). Change blindness. *Trends in Cognitive Sciences*, 1, 261–7. doi: 10.1016/S1364–6613(97)01080–2

Simons, D.J. and Levin, D.T. (1998). Failure to detect changes to people during a real-world interaction. *Psychonomic Bulletin & Review*, 5, 644–9.

Simons, D.J. and Rensink, R.A. (2003). Induced failures of visual awareness. *Journal of Vision*, 3(1). doi: 10.1167/3.1.i

Smither, J.A. and Torrez, L.I. (2010). Motorcycle conspicuity: Effects of age and daytime running lights. *Human Factors*, 52(3), 355–69. doi: 10.1177/0018720810374613

Summala, H., Pasanen, E., Räsänen, M. and Sievänen, J. (1996). Bicycle accident and drivers' visual search at left and right turns. *Accident Analysis & Prevention*, 28, 147–53.

Thakral, P.P. and Slotnick, S.D. (2010). Attentional inhibition mediates inattentional blindness. *Consciousness & Cognition*, 19, 636–43. doi: 10.1016/j.concog.2010.02.002

Theeuwes, J. (1992). Perceptual selectivity for color and form. *Perception & Psychophysics*, 51, 599–606.

Thomson, G.A. (1980). The role frontal motorcycle conspicuity has in road accidents. *Accident Analysis & Prevention*, 12(3), 165–78. doi: 10.1016/0001–4575(80)90015–9

Todd, J., Fougnie, D. and Marois, R. (2005). Visual short-term memory load suppresses temporo-parietal junction activity and induces inattentional blindness. *Psychological Science*, 16, 965–72. doi: 10.1111/j.1467–9 280.2005.01645.x

Treisman, A.M. and Gelade, G. (1980). A feature-integration theory of attention. *Cognitive Psychology*, 12, 97–136.

Underwood, G. (2009). Cognitive processes in eye guidance: Algorithms for attention in image processing. *Cognitive Computation*, 1(1), 64–76. doi: 10.1007/s12559–008–9002–7

Van Elslande, P. and Faucher-Alberton, L. (1997). When expectancies become certainties: A potential adverse effect of experience. In Rothengatter, T. and Carbonell Vaya, E. (eds), *Traffic and Transport Psychology: Theory and application*, 147–59. Oxford: Elsevier.

Van Elslande, P., Fournier, J.-Y. and Jaffard, M. (2012). In-depth analysis of road crashes involving powered two-wheelers vehicles: Typical human functional failures and conditions of their production. *Work*, 41, 5871–3. doi: 10.3233/WOR-2012–0978–5871

Van Wert, M.J., Horowitz, T.S. and Wolfe, J.M. (2009). Even in correctable search, some types of rare targets are frequently missed. *Attention, Perception, & Psychophysics*, 71(3), 541–53. doi: 10.3758/app.71.3.541

Velichkovsky, B.M., Dornhoefer, S.M., Kopf, M., Helmert, J. and Joos, M. (2002). Change detection and occlusion modes in road-traffic scenarios. *Transportation Research Part F: Traffic Psychology & Behaviour*, 5(2), 99–109. doi: 10.1016/S1369–8478(02)00009–8

Volkmann, F.C., Riggs, L.A., Moore, R.K. and White, K.D. (1978). Central and peripheral determinants of saccadic suppression. In Senders, J.W., Fisher, D.F. and Monty, R.A. (eds), *Eye Movements and the Higher Psychological Functions*, 35–54. Hillsdale, NJ: Lawrence Erlbaum.

Walker, G.H., Stanton, N.A. and Salmon, P.M. (2011). Cognitive compatibility of motorcyclists and car drivers. *Accident Analysis & Prevention*, 43, 878–88. doi: 10.1016/j.aap.2010.11.008

Wayand, J.F., Levin, D.T. and Varakin, D.A. (2005). Inattentional blindness for a noxious multimodal stimulus. *American Journal of Psychology*, 118, 339–52.

Wells, S., Mullin, B., Norton, R., Langley, J., Connor, J., Lay-Yee, R. and Jackson, R. (2004). Motorcycle rider conspicuity and crash related injury: Case-control study. *BMJ*, 328(7444), 857. doi: 10.1136/bmj.37984.574757.EE

White, R.C. and Aimola Davies, A. (2008). Attention set for number: Expectation and perceptual load in inattentional blindness. *Journal of Experimental Psychology: Human Perception & Performance*, 34, 1092–107. doi: 10.1037/0096–1523.34.5.1092

Williams, M.J. and Hoffman, E.R. (1979a). Conspicuity of motorcycles. *Human Factors*, 21(5), 619–26.

Williams, M.J. and Hoffman, E.R. (1979b). Motorcycle conspicuity and traffic accidents. *Accident Analysis & Prevention*, 11, 209–24.

Wolfe, J.M. (1999). Inattentional amnesia. In Coltheart, V. (ed.), *Fleeting Memories: Cognition of Brief Visual Stimuli*, 71–94. Cambridge, MA: MIT Press.

Wolfe, J.M. (2007). Guided search 4.0: Current progress within a model of visual search. In Gray, W.D. (ed.), *Integrated Models of Cognitive Systems*, 99–119. New York, NY: Oxford University Press.

Wolfe, J.M., Horowitz, T.S. and Kenner, N.M. (2005). Rare items often missed in visual searches. *Nature*, 435(7041), 439–40. doi: 10.1038/435439a

Wolfe, J.M., Horowitz, T.S., Van Wert, M.J., Kenner, N.M., Place, S.S. and Kibbi, N. (2007). Low target prevalence is a stubborn source of errors in visual search tasks. *Journal of Experimental Psychology: General*, 136(4), 623–38.

Young, J.E., Klosko, J.S. and Weishaar, M.E. (2003). *Schema Therapy: A Practitioner's Guide*. New York, NY: Guilford Press.

Chapter 3

Mechanisms Underpinning Conspicuity

Geoff Underwood

How do road users scan the roadway when deciding to make a manoeuvre, and what governs whether they will see another road user whose presence might influence their decision? This is a special instance of a general question that has long occupied cognitive psychologists: the question of 'what attracts attention?'.

To traffic psychologists and to cognitive psychologists alike, there are two categories of reply to the question: the internal, schema-driven control of attention determined by the road user's driving goals and situation awareness, and the external, event-driven control of attention whereby some object in the environment captures attention by its unexpected activity. This dichotomy is sometimes referred to as the distinction between the endogenous and exogenous control of attention, and can be characterized by the distinction between a driver knowing that he should stop before entering a junction, in order to assess the oncoming traffic, and a driver noticing a fast-moving vehicle that requires evasive action. In one case the driver's understanding of traffic dynamics requires inspection of the roadway, and in the second case an event in the roadway captures the driver's attention. The dichotomy is too simplistic of course, and in many situations our attention is distributed according to our knowledge of where important events are likely to occur as well as being drawn to unusual or sudden-onset events such as a pedestrian walking out from behind a parked vehicle. On some occasions we do not inspect the roadway as we should, perhaps being distracted by another event, and other occasions we might rigidly scan the roadway locations where our situation awareness leads us to anticipate oncoming traffic, to the exclusion of a pertinent event in an unexpected part of the roadway. As an example of this interaction, imagine walking across an open pedestrian space – a city square, perhaps – while engaged in conversation with someone either on a mobile phone or who is walking with you. If something bizarre happened, would your attention be drawn to it? Suppose now a brightly dressed clown on a unicycle crossed your path (he was there to advertise a circus coming to town). Would your attention be attracted to the clown? If attention had been inevitably drawn by bizarre exogenous events, then the clown would be noticed. When navigating the environment, attention cannot be locked completely on to internal events or even a conversation, because the route must be maintained and our progress monitored, so there must be some notice taken of environmental events. This situation has actually been studied, in an experiment conducted by Hyman et al. (2010), who questioned walkers after they had crossed a pedestrian plaza on a university campus in the presence of a unicycling clown. The answer is that individuals using a phone were half as likely to notice the clown relative to individuals not using a phone. This example of inattentional blindness, perhaps best known from the example of the undetected 'gorilla' amongst ball players (Simons and Chabris, 1999), illustrates that events that might be expected to be attentionally demanding in some circumstances can be invisible if attention is elsewhere. Endogenous and exogenous mechanisms are more likely to interact than to be in opposition in an 'either/or' fashion, but the distinction serves here to indicate our concern with the emphasis on external events in the control of a driver's attention. Essentially the debate concerns the ways in which the prior allocation of attention can override bottom-up exogenous attentional capture and top-down endogenous attentional guidance. Evidence has suggested that both the goal-based searching for objects and the effects of unexpected conspicuous objects are influenced by the search goals from the recent past (for a review, see Awh et al., 2012). Both of these processes are relevant to driving of course, with the unexpected appearance of another road user being of potential importance and in need of processing, just as the volitional monitoring of the roadway is important at all times. We shall return to this distinction after considering the exogenous capture of attention in more detail.

Cognitive Conspicuity and Visual Conspicuity

A layman's reply to the question of what attracts our attention might be to say that attention goes to interesting parts of the scene, but this diverts the question firstly to one of what makes part of a scene 'interesting', and secondly, how it could be that the viewer knew that it was interesting *before* moving their eyes to inspect it. How could it be processed to the point where it was known to be of interest without it being inspected? We move our eyes to inspect objects in order to bring the image of the object onto area of the retina known as the fovea. This is a relatively small area (subtending approximately 2 degrees of visual angle)[1] but where there is the greatest density of retinal receptor cells and where inspection will be most effective. Moving our eyes to an interesting part of the scene is therefore necessary if it is to gain the most detailed analysis. There is some dispute over whether we move our eyes *after* detecting the presence of something of meaningful interest that is not under foveal scrutiny, with mixed evidence for the detection of interesting objects prior to fixation (for a review, see Underwood, 2009). Evidence of the processing of meaningful objects prior to foveal inspection would take the form of viewers detecting a sauce bottle in a picture of a bathroom

---

1   To have an impression of the size of your fovea try this exercise: with your arm extended, look at your thumbnail – it will occupy most of your fovea at this distance. This is, of course, a generalization, and the actual area subtended by your thumbnail will depend upon its size as well as the length of your arm, but it does illustrate the small area of the fovea relative to the total visual field.

scene, for example, and moving their eyes to the bottle earlier than if it had been a shampoo bottle in the same place, but support for such processing of cognitively conspicuous objects is sometimes found and sometimes not. This example of an anomalous object – a sauce bottle replacing a shampoo bottle, in a scene containing a soap dish and toothpaste for example – is an extreme illustration of cognitive conspicuity, or an object that is conspicuous by virtue of being unexpected in that location. Expectancy is fundamental to cognitive conspicuity, in that the viewer will have developed an understanding of what is probable in any known specific situation, and understanding of an object's probability of appearance will be based on experience. A sauce bottle on the bath side is an anomaly – it is in the wrong place by functionality – and equivalents can be imagined for roadway anomalies such as the unicycling clown (Hyman et al., 2010), and less extreme anomalies involving vehicles that are unexpected. Motorcycles have a lower probability of occurrence relative to cars and therefore can be described as being cognitively inconspicuous. The discussion of whether 'look but fail to see' motorcycle crashes are related to cognitive inconspicuity will be continued elsewhere in this volume (see, for example, the chapter by Rogé and Vienne). While there is a debate about the early detection of anomalous objects in visual scenes, however, there is clearer support for the notion of the early processing of visually conspicuous objects.

Visual Conspicuity and Visual Saliency

The terms *visual conspicuity* and *visual saliency* are used interchangeably here and in the traffic psychology literature generally, but not in the vision science community. Generally, saliency (or salience) is the overall term that encompasses all of the visual characteristics that distinguish an object from its surroundings – the characteristics that make it stand out. The specific characteristics themselves are measured along separate dimensions, and these are measures of conspicuity such as brightness, colour and movement. Thus, we can talk of brightness conspicuity, colour conspicuity, movement conspicuity and so on, and each of these will add up to the total saliency of the object. The way in which the conspicuity channels are aggregated to generate a saliency map of the image is best illustrated by considering one of vision science's models of saliency. There are many models that attempt to account for the relationships between attention and visual saliency, and the most influential of these is a model proposed by Laurent Itti and Christof Koch (2000). The influence of this model stems from two of its attributes – completeness and accuracy in predicting behaviour. Several other models are described in the special issue of the journal *Cognitive Computation* edited by Taylor and Cutsuridis (2011), but it is sufficient here to consider just one model, in view of its longevity and its success in accounting for the way that attention moves around the visual world.

Briefly, the Itti and Koch (2000) saliency map model predicts the initial eye fixations on a scene by detecting areas of difference in a number of conspicuity

Input frame:

Figure 3.1   A photograph of a natural scene with variations in colour, brightness and orientation

Note: The Itti & Koch (2000) model predicts where our eyes should be attracted when we first look when first look at pictures such as this.

Source: Reprinted by permission. *Cognitive Computation.* 2009, Springer: 67.

channels including colour and orientation, aggregating these peaks of difference into a single saliency map, and then directing the first fixation to the most salient point, the second fixation to the next most salient point and so on. We can illustrate how the model works by considering a viewer's inspection of the photograph in Figure 3.1.

It is important to stress that the early analysis is a purely bottom-up process, with the meaning of the objects being totally irrelevant. There may be some interest in the question of what it is on the boat that the bystanders on the quayside are finding so engrossing, or in the question of where the photograph was taken or who the bystanders are, but a bottom-up analysis takes account only of the visual components of the image.

Analysis in Variations in:

Intensity   Colour   Orientation

Figure 3.2   The image in Figure 3.1 is analysed for variations in a number of purely visual dimensions, with the intensity, colour and orientation channels being illustrated here

Note: The three images here are separate analyses of variations in these channels, with intensity peaks corresponding to objects in the centre right (people on the quayside) of the original photograph, major colour variations in the centre left (the fishing boat), and changes in orientation in centre right (people, especially a hat in the extreme upper right quadrant) and lower centre (the edge of the quay).

Source: Reprinted by permission. *Cognitive Computation.* 2009, Springer: 67.

Variations in the visual components are computed using a centre-surround principle to identify local contrasts between a specific area and its immediate surround. Each calculation in this process of feature extraction is done at six scale sizes, where the size of centre and the size of the surround vary. Separate channels initially analyse intensity (one channel), colour (two channels – one for red/green variation and one for blue/yellow variation), and orientation (four channels – 0, 45, 90 and 135 degrees) making seven feature maps, and these are computed at the six different scales to make

Raw Saliency Map

Figure 3.3 **The values of the peaks of variations from the separate analyses (Figure 3.2) are aggregated into a combined raw saliency map**

Source: Reprinted by permission. *Cognitive Computation.* 2009, Springer: 67.

42 feature maps. Figure 3.2 shows a simplified output from the computation applied to the photograph in Figure 3.1, with the combined conspicuity maps for intensity, colour and orientation giving an indication of how these variations map onto the colour photograph.

The 42 feature maps are combined to create individual conspicuity maps using a process of normalization. The measures of variation for the different features are not equivalent and so they need to rendered along a comparable dimension. The problem of equivalence is handled by a process of Gaussian normalization. This problem is one of combining a variation of, say, 10 degrees in the orientation channel with, say, a 5 per cent increase in contrast. These are measures of discontinuity, and the model weights them by normalization so that the range of variation over the whole image is taken into account, and relatively large variations weighted along comparable dimensions. The resultant conspicuity maps (Figure 3.2) highlight the most significant discontinuities for a specific visual attribute. A similar process is applied to the conspicuity maps to generate a combined raw saliency map (Figure 3.3). The objects appearing in the map can now be seen to correspond to the objects in the original input frame.

The combined raw saliency map is used to guide attention around the image. The area with the greatest combined discontinuity value – the highest saliency peak – attracts the first eye fixation using a 'winner-takes-all' algorithm. This processes the highest peak and uses it to direct the first saccadic eye movement to

**Figure 3.4   The Itti & Koch (2000) model predicts that the order of eye
fixations will match the rank order of saliency peaks derived
from the raw saliency map**

Note: These peaks are shown here as predicted eye movements that start with a fixation
on the white hat in the upper right quadrant. The second fixation should be on an area of
quayside where someone is seated, the third on the boundary between the fishing nets and
the side of the boat, and so on. If fixation should always be attracted to the most salient area,
then when the second fixation is made the most salient area away from fixation becomes
the recently fixated white hat. It would be unproductive to move the gaze directly back to
area 1, and then back to area 2, and so on, as required by the simple algorithm of invariably
moving to the highest available saliency peak, and so the model assumes that a principle of
'inhibition of return' operates. Once an area has been inspected the input from this saliency
peak is suppressed, to allow attention to move around the image.

that point. After fixation the eyes move to the next highest peak, and at the same
time an 'inhibition of return' process is applied to suppress the previous highest
peak. Without this process attention would flip between the two highest peaks
and would not be able to move over the image. In Figure 3.4, the six most salient
regions are identified by numbers that also correspond to the first six predicted
fixations. Without inhibition of return, the focus of attention would bounce from
inspection of the white hat (upper right quadrant) and the quayside in the central
right of the photograph and then back again repeatedly, and peak 3 (nets and boat

edge) would not be inspected at all. With this process in place, attention is allowed to move around the image to each saliency peak in turn.

Tests of the Saliency Maps Model: Evidence from Eye Movements

Visual anomalies in otherwise uniform displays are detected very easily. This is the so-called 'pop-out' effect where one item amongst an array of other items will be seen without effort: on a kitchen table, for example, a carrot will stand out amongst a set of mushrooms, or in a garden a single red rose will be conspicuous in a bed of white roses. This pop-out effect has been investigated in psychology laboratories for some time now, and is known to be robust and with well-established characteristics (see, for example, Treisman and Gelade, 1980). In the laboratory, the effect can be demonstrated by displaying a set of short lines all oriented in the same direction except for a singleton oriented at 90 degrees to the others. Or the lines might be all one colour with a singleton in another colour. In both cases the singleton stands out from the array – it is conspicuous. A less conspicuous target would be detected more slowly as the number of homogenous distractors is increased, but a conspicuous target is resistant to this effect of the varying number of non-targets. This effect provides a simple test of the saliency map model, and Itti and Koch (2000) used colour pop-out and orientation pop-out as a first empirical test. An array of small green rectangles provided the background, with one of the rectangles replaced by a red version. This singleton was detected successfully by the model, with the number of distractors (green rectangles) having no effect on the detection probability, just as it would with human observers. A similar effect was seen with orientation as the key variable: rectangles aligned together (all pointing top left to bottom right, for instance) provided a distinctive background for a singleton that pointed top right to bottom right. For both the model and people this conspicuous singleton stands out and is easily detected, regardless of the number of distractors.

The model also performs well in comparison with people inspecting pictures of scenes. Parkhurst et al. had participants look at a variety of images while their eye fixations were recorded. No specific task was presented to the participants, who were asked to 'look around at the images' (2002: 112) for 5 secs. The question asked in their study was whether people tended to look at the high-saliency areas of the pictures earlier than other areas, and they did. The images were photographs of indoor and outdoor scenes as well as computer-generated fractals, and the first few fixations were recorded during this 'free viewing' task. Using the inspection of abstract and depictive paintings, Fuchs et al. (2011) also used a free-viewing task to confirm the conclusion that eye fixations are drawn to highly salient regions of an image. It is difficult to anticipate the perceived purpose of the free-viewing task, however, with participants wondering whether they were to be presented with some kind of surprise recognition test at the end of the viewing exercise, and this uncertainty in the minds of the viewers may have clouded the results by inducing a particular inspection strategy. Whatever the perceived purpose of the task, the first few fixations tended to be focused on the high saliency areas in the images in these studies, and we have found that when a specific encoding task is set for the participants then fixations are again attracted towards high-saliency areas.

In general, experiments that use memory tasks lend support to the saliency map model, but those using search tasks do not. In experiments using photographs of indoor scenes we had participants first look at pictures for a few seconds, and then look at a second set of pictures to say which of the pictures in the second set had been seen in the first set (Underwood et al., 2006; Underwood and Foulsham, 2006). Their eye movements were recorded while they looked at the first set of pictures, and early fixations tended to land on high-saliency areas of the pictures. The same pictures were then used in search tasks, with new participants. In one experiment the target was the general category of an example of a piece of fruit (Underwood et al., 2006), and in the second case the target was a well-specified but inconspicuous object (Underwood and Foulsham, 2006). In these experiments fixations were not associated with the saliency peaks in the photographs, thereby challenging the generality of the saliency map hypothesis. Fuchs et al. (2011) also found that the influence of saliency was moderated when viewers searched for a specific object rather than when they looked at the same scenes in a free-viewing task. The distinction between memory tasks and search tasks can be related to differences between bottom-up exogenous attentional capture and top-down endogenous attentional guidance. In memory tasks viewers look around the image for memorable features that will help them in the recognition phase, and they will allow highly conspicuous features to attract their attention. In search tasks, on the other hand, the viewers have a specific target in mind and will guide their attention to likely locations in the scene. In this case the endogenous guidance can resist the bright, colourful distractors and focus on the target. When I search for my car keys on my desktop, my attention is not captured by brightly coloured pens or notepaper and only the potential locations of the keys are inspected.

The conclusion that saliency guides fixations in a memory task can also be challenged. When experts in a domain look at pictures from their domain, their fixations are guided by their knowledge rather than by the conspicuity of the objects in the scene (Humphrey and Underwood, 2009; Underwood et al., 2009). In these experiments we had two groups of participants look at pictures that were from their own domain of interest and expertise, or from the domain of the other group. Students were recruited from a university engineering department, and from a university course concerned with American history. They were shown pictures of engineering installations and pictures of memorabilia from the American Civil War, in the same memory task used previously. Both groups were shown both kinds of pictures, and the results supported the saliency map hypothesis for pictures from an unfamiliar domain, but not when the pictures matched the participants' interests. Engineering students did not look at the brightly coloured areas of engineering installations, and instead followed the functionality of the structures. Conversely, the American Studies students characteristically did look at the conspicuous features and made no attempt to guess the purpose of the structures. The opposite was the case for Civil War pictures, with engineers looking at the colourful features and the American Studies students concerned, perhaps, with the provenance of the objects depicted. Saliency is most effective, we might conclude, when the scene has limited meaning, or at least, no especial meaning for the viewer. Searching for specific objects is not strongly disrupted by the presence of high-saliency non-target areas, but looking at neutral scenes – those without special meaning – does induce fixation on salient areas. The importance of familiarity has also been demonstrated by Huestegge and Radach (2012) who had their participants search a set of shelves for a specific fruit juice container. Saliency, as assessed by the Itti and Koch (2000) algorithm, was an effective factor in this search task, but only for products declared by the participants as being unfamiliar to them. This study also provides a caveat to our conclusion that saliency has no role in search tasks – when searching for a well-specified target that is shown immediately prior to the display, then saliency will influence the pattern of search for unfamiliar objects.

The saliency map model works well in describing performance in some situations but not all, and further limitations to the success of the model become apparent when the pictures in a memory experiment contain socially significant objects. These images can be expected to induce the kind of special interest seen in the experiments using participants with special domains of interest or familiarity. Humphrey and Underwood (2010) using pictures with people in them, and Humphrey et al. (2012) using pictures of emotional scenes, also found that the saliency map did not make good predictions about what viewers would inspect. We can now summarise the advantages of conspicuity for vehicles that need to be seen by other road users, but first there is a cautionary note about types of conspicuity.

Conspicuity and Concealment

If distinguishing an object from its background can be regarded as forming a continuum, then concealment and camouflage can be considered to be at the opposite end of the continuum from conspicuity. In one case the object will stand out from its background and in the other it will blend and become difficult to discriminate from surrounding objects. Camouflage is of benefit for animals that need to avoid the attention of predators, and of course this includes military concealment.

The butterfly in Figure 3.5 is both camouflaged (left image) and conspicuous (right image), and this illustrates how detectability can vary according to background alone. Animals that avoid detection by blending with their backgrounds are said to employ *crypsis* camouflage, by using patterned markings that break up their outlines, by eliminating shadows by pressing their bodies to the ground or by actively changing their skin colour, for example. Examples include animals with body colour and pattern that break up or disrupt their outline (for example, leopards) or that change their body colour to blend with their current background (for example, chameleon, octopus). This is distinguished from *mimesis* camouflage,

Figure 3.5   **Two images of the same butterfly photographed from different angles – In the image on the left the butterfly blends with its background (leaves) but on the right it is more conspicuous**

whereby the animal has evolved to resemble something else such as a leaf (Figure 3.5). Other examples of this mimicry include harmless animals that resemble something more dangerous (for example, hoverfly/wasp). Most camouflage effects are broken when the animal moves but with a third form of camouflage a *dazzle* pattern only has its effect when the object moves. The stripes of a zebra make it difficult for a predator to identify one animal when several move together, and a similar principal has been used for military purposes. Warships, such as those shown in Figure 3.6, are quite conspicuous when stationary, but the disruptive patterns are thought to make the estimation of speed and direction more difficult. This is important, of course, if it induces errors in an enemy's aiming of weapons. Although military analyses of the efficacy of dazzle camouflage are not compelling, there is recent evidence to suggest that dazzle patterns can

Figure 3.6   **Two examples of First World War vessels painted with dazzle camouflage patterns – Upper image is USS Leviathan and lower image is HMS London**

Source: Wikimedia Commons. See http://commons.wiki media.org/wiki/

indeed influence the estimation of speed, and this is of some consequence for roadway applications.

To evaluate the possible effect of dazzle camouflage on the perceived speed of movement of objects, Scott-Samuel et al. (2011) had viewers watch two patches displayed on a computer screen and to estimate which of them moved more quickly. One of the patches was a comparison stimulus, in that it showed no patterning, and the test stimuli had straight vertical or horizontal bands or a zigzag pattern or a chequered pattern. Both the comparison and test stimuli appeared very briefly on the screen on each of several trials. The patterns moved quickly (20 deg/sec) or slowly (3.3 deg/sec) and were shown under high contrast or low contrast luminance conditions. There were no apparent effects for low speed discriminations, and none for high speed, low contrast stimuli, but fast moving high contrast patterns did shown an effect. The zigzag pattern and the chequered pattern were both perceived as moving more slowly than the plain stimulus under these viewing conditions. This underestimate of speed of stimuli that were patterned in a way similar to that used in military applications would encourage the use of dazzle camouflage, and Scott-Samuel et al. (2011) point out that the disruptive effect on speed perception operates best for a vehicle at a distance of 70 m travelling at 90 kph. At this distance a vehicle's speed would be underestimated by about 7 per cent, leading observers to believe that it was travelling at around 84 kph and, therefore, that they would have more time to enter the junction before its arrival.

If a motorcycle were to be painted with graphics that made it conspicuous while stationary but that resembled the dazzle patterns that produce the effects seen here, then there would be a counter-productive effect of rendering the vehicle as more vulnerable to incorrect estimations of its approach speed. Examples of potential 'dazzle camouflage' graphics are shown in Figure 3.7. When judging whether to enter a main road from a side road, drivers will inspect the traffic on the main road and if there is an approaching vehicle they will estimate the time

Figure 3.7   Examples of motorcycle graphics that may induce a speed under-estimation error on the part of other road users

Source: Wikimedia Commons. See http://commons.wikimedia.org/wiki/

available to make the manoeuvre to enter the road. That estimate is made partly on the distance of the approaching vehicle and partly on its speed, and if its speed is underestimated a faulty decision is possible. Dazzle graphics on a motorcycle (for example Figure 3.7) may render it more attractive in the showroom and more conspicuous when stationary, but may have the effect of inducing faulty decisions by road users estimating its speed. The difference between the actual speed and the misestimated speed may be sufficient to lead a road user to believe that there is sufficient time to enter the junction when in fact the motorcycle would arrive during the manoeuvre rather than afterwards.

Summary: What Saliency Models Tell Us About Inspecting the Scene

Objects such as motorcycles can appear to be more visible than their surroundings by contrasts in a number of conspicuity variables such as colour, orientation and brightness. The sum of all of the available conspicuity measures result in a saliency value, and the Itti and Koch (2000) saliency map model provides a good description of how attention is attracted to objects according to the way that they stand out against their backgrounds. According to the model, a bright yellow vehicle (and rider) would stand out against a black/grey roadway environment better than a black motorcycle with a rider in black clothing, but against a brightly coloured urban environment the same contrast might not apply. The model successfully predicts that highly salient objects attract earlier inspection, faster decision-making, and more cautious decision-making, when road users are asked about entering a road junction with an approaching motorcycle (Underwood et al., 2011). The empirical chapters in this book develop this basic finding and describe the generality of the effects of high saliency in determining the extent to which a motorcycle and its rider can be made more visible to other road users.

There are some limitations on the success of the saliency map hypothesis. The model works very well in predicting the initial inspection of an image when viewers are given no specific instructions, or when they engage in a task requiring them to encode the image in preparation for a later recognition test. Familiarity and social significance moderate these conclusions, with saliency having a greater influence with unfamiliar and non-emotive scenes, and less well when viewers search for a specific object that is described (for example, 'search for a piece of fruit') rather than a target object displayed immediately prior to the search scene. When attention is guided in a top-down endogenous fashion with a specific target object in mind, then conspicuous objects are no more attention-capturing than any others. If the viewer is inspecting the scene more generally, then conspicuous objects will attract attention early. Expertise moderates the effects of conspicuity, and this may transfer to the road environment by road users with specific interests and detailed knowledge behaving differently from other road users. Motorcycle riders with special interests – hobby riders or motorcycle club members perhaps – may be less affected by a conspicuous motorcycle than a car driver with no particular interests in motorcycles.

References

Awh, E., Belopolsky, A.V. and Theeuwes, J. (2012). Top-down versus bottom-up attentional control: A failed theoretical dichotomy. *Trends in Cognitive Sciences*, 18, 437–43.

Fuchs, I., Ansorge, U., Redies, C. and Leder, H. (2011). Salience in paintings: Bottom-up influences on eye fixations. *Cognitive Computation*, 3, 25–36.

Huestegge, L. and Radach, R. (2012). Visual and memory search in complex environments: Determinants of eye movements and search performance. *Ergonomics*, 55, 1009–27.

Humphrey, K. and Underwood, G. (2009). Domain knowledge moderates the influence of visual saliency in scene recognition. *British Journal of Psychology*, 100, 377–98.

Humphrey, K. and Underwood, G. (2010). The potency of people in pictures: Evidence from sequences of eye fixations. *Journal of Vision*, 10(10):19, 1–10.

Humphrey, K., Underwood, G. and Lambert, T. (2012). Salience of the lambs: A test of the saliency map hypothesis with pictures of emotive objects. *Journal of Vision*, 12(1):22, 1–15.

Hyman, I.R., Boss, S.M., Wise, B.M., McKenzie, K.E. and Caggiano, J.M. (2010). Did you see the unicycling clown? Inattentional blindness while walking and talking on a cell phone. *Applied Cognitive Psychology*, 24, 597–607.

Itti, L. and Koch, C. (2000). A saliency-based search mechanism for overt and covert shifts of visual attention. *Vision Research*, 40, 1489–506.

Parkhurst, D., Law, K. and Niebur, E. (2002). Modelling the role of salience in the allocation of overt visual attention. *Vision Research*, 42, 107–23.

Scott-Samuel, N.E., Baddeley, R., Palmer, C.E. and Cuthill, I.C. (2011). Dazzle camouflage affects speed perception. *PLoS ONE*, 6(6), 1–5.

Simons, D.J. and Chabris, C.F. (1999). Gorillas in our midst: Sustained attentional blindness for dynamic events. *Perception*, 28, 1059–74.

Taylor, J.G. and Cutsuridis, V. (2011). Saliency, attention, active visual search and picture scanning. *Cognitive Computation*, 3, 1–332.

Treisman, A.M. and Gelade, G. (1980). A feature-integration theory of attention. *Cognitive Psychology*, 12, 97–136.

Underwood, G. (2009). Cognitive processes in eye guidance: Algorithms for attention in image processing. *Cognitive Computation*, 1, 64–76.

Underwood, G., Foulsham, T. and Humphrey, K. (2009). Saliency and scan patterns in the inspection of real-world scenes: Eye movements during encoding and recognition. *Visual Cognition*, 17, 812–34.

Underwood, G. and Foulsham, T. (2006). Visual saliency and semantic incongruency influence eye movements when inspecting pictures. *Quarterly Journal of Experimental Psychology*, 59, 1931–49.

Underwood, G., Foulsham, T., van Loon, E., Humphreys, L. and Bloyce, J. (2006). Eye movements during scene inspection: A test of the saliency map hypothesis. *European Journal of Cognitive Psychology*, 18, 321–42.

Underwood, G., Humphrey, K. and van Loon, E. (2011). Decisions about objects in real-world scenes are influenced by visual saliency before and during their inspection. *Vision Research*, 51, 2031–8.

PART II
Case Studies Focusing on Visual Saliency and Conspicuity Treatments

Chapter 4

How Conspicuity Influences Drivers' Attention and Manoeuvring Decisions

Geoff Underwood, Editha van Loon and Katherine Humphrey

Introduction

How do conspicuous vehicles attract the attention of other road users, and do they capture the attention of other road users more readily than less conspicuous vehicles? Two studies address these questions by having drivers and riders assess whether the traffic in a roadway scene would allow or discourage them from making one of two manoeuvres. In the first study they decided whether it would be safe to enter a major road from a side road at a T-junction, and in the second study they decided whether it would be safe to move into the outer lane of a two-lane highway in order to overtake a slower vehicle or in preparation for making a turn at a junction. In both studies there were vehicles sometimes present, and these could be conspicuous or not. For several decades, safety campaigners have pointed to greater conspicuity being associated with reduced crash risk for vehicles, with epidemiological and quasi-experimental studies providing a consistent conclusion (for example, Olson et al., 1979, 1981; Thomson, 1980; Zador, 1985; Elvik, 1993; Yuan, 2000).

Accident researchers often use the term 'conspicuity' to indicate a part of the scene that stands out from the background (such as high-visibility clothing), whereas vision scientists usually use the term 'saliency' where different types of conspicuity (provided by variations in brightness, intensity, orientation and so on) each generate separate conspicuity maps of the image and which are then combined to form an overall saliency map of the scene (see Chapter 4 on the assessment of conspicuity and saliency values). We will use the two terms interchangeably here, although the assessment of conspicuity is done with reference to the overall saliency map of the photographs used in the two experiments, and so 'saliency' is the more correct term. The conclusion that conspicuous vehicles should be more noticeable by other road users is supported by saliency map models of visual attention that predict the direction of our eyes when we first look at a scene. Conspicuous regions of a scene are predicted to gain early inspection by virtue of the early acquisition of information about objects that are easily discriminated from their surroundings. Changes in colour and brightness identify an object as being visually salient, and these are the objects that are said to attract our attention. The advantage of salient over non-salient regions is considered to be temporary and most apparent during the early inspection of the scene, however, and after the first few eye fixations then top-down conceptual processes will dominate and the viewer's attention will be captured by meaningful objects. The two studies described here assess the saliency map model as a predictor of sequences of fixations when drivers and riders inspect photographs of roadway scenes in preparation for a decision about a possible manoeuvre. We recorded the locations of eye fixations of viewers when they looked at images in order to decide whether it would be safe to make a specified manoeuvre: to determine firstly whether the decision depended on the conspicuity of a vehicle in the picture and secondly when any effects occurred. Effects of conspicuous vehicles could occur entirely during its inspection, or they might occur prior to inspection if a conspicuous vehicle can attract attention in the way predicted by saliency map models.

The most explicit and fully formalised models of the saliency map model was published and made generally available by Itti and Koch (2000). Their algorithm measures the saliency of separate regions of image objectively, by the identification of peaks in the distribution of colour, intensity and orientation. These are the separate conspicuity maps that are combined to form an overall saliency map. The distribution of saliency values generates predictions as to where attention should be directed and forms the basis of a model of visual attention. For each of the visual characteristics a separate conspicuity map is first computed by searching for change relative to adjacent regions. These maps are then combined to find saliency peaks, with a change in any of the three characteristics resulting in an increase in the saliency value assigned to that region of the image. The analysis of low-level visual information is also central in models of scene processing such as those of Henderson et al. (1999) and Findlay and Walker (1999). Early visual processes determine the initial fixations during picture inspection in these models, and the meaning of objects and events depicted can be appreciated only after the first few fixations.

The saliency map model of visual attention has been supported by the results from search tasks with simple geometric displays of targets and distractors (for example, Nothdurft, 2002; Lamy et al., 2004), but the model also predicts eye fixations on natural images. Parkhurst et al. (2002) and Fuchs et al. (2011) presented a range of images, including photographs of real-world scenes and paintings. The saliency values of regions in each image provided a good prediction of the order of eye fixations during the few seconds available for inspection. We have also established the relationship between task purpose and image saliency in experiments that used the same images and different tasks (Underwood et al., 2006; Underwood and Foulsham, 2006; Foulsham and Underwood, 2007). When instructed to inspect scenes in preparation for a recognition test, then fixations followed the predictions of the saliency map model, with conspicuous objects attracting early fixations. But when instructed to decide whether the scene contained a specific target object, however, a highly salient non-target distractor object was not fixated in preference to a less salient target.

Saliency can be seen to determine the capture of attention, but only when there is no object-based purpose to the inspection, as there is when searching for a specific object. The absence of attentional capture by salient objects has also been demonstrated in change detection experiments (Stirk and Underwood, 2007; Underwood et al., 2008). A single object was changed between two images of the same scene in these experiments, and the object was either very conspicuous when changed or it was inconspicuous but the saliency of the changed item did not influence detection. These experiments looking at eye movements made during the inspection of photographs provide some support for the saliency map hypothesis, but the early attraction of attention depends upon the task presented. The objects in the scene are also important to the early attraction of attention, and when a scene depicts people then the saliency of other objects has reduced influence (Henderson et al., 2007; Humphrey and Underwood, 2010). The purpose of inspection and the meaningful components of the scene are taken into account by more recent models of real-world scene perception that do not rely exclusively on bottom-up processes (Navalpakkam and Itti, 2005; Torralba et al., 2006), but saliency map models continue to attract criticism. For example, Tatler (2007) has described the dominance of the central fixation tendency in scene perception, as an example of how scene-independent biases may produce consistent patterns that have little or nothing to do with bottom-up saliency or top-down conceptual guidance.

Would a conspicuous vehicle attract early attention when a viewer inspects the scene with a specific purpose? Or would the high-level cognitive purpose of inspection override the low-level visual influences of conspicuity? Two experiments investigated the influence of saliency when road users assessed the risks involved in making manoeuvres when other vehicles were present.

Method

Participants

The volunteers in the two studies here were 50 car drivers (mean age 36.7 years, range 20–64 years) and 27 motorcycle riders (mean age 37.5 years, range 20–66 years), each with a minimum of two years experience since obtaining their licence. They all had normal or corrected-to-normal vision, and were paid an inconvenience allowance.

Materials and Apparatus

A set of 160 high-resolution digital photographs were prepared as stimuli, taken using a 9MP digital camera and edited using Adobe Photoshop. Of these, 120 were taken from a driver's perspective through a side window in a car that had pulled up to a T-junction. These photographs were used in Study 1, in which a judgement was required about whether it would be safe to move into the main roadway,

a)   b)

**Figure 4.1   T-Junction pictures from a car's side windows. Traffic
approaching from the a) left or from the b) right**

joining either to the left or to the right. All roads were dual-carriageways ('urban motorway'), so both lanes of traffic were always travelling in the same direction. To prevent the participants from anticipating the location of the traffic and then moving their eyes before each picture appeared, the traffic would sometimes be coming from the left, and sometimes from the right (see Figure 4.1).

Of the T-junction pictures, 10 had motorcycles approaching at mid-distance, 10 had cars approaching at mid-distance, 40 had cars or motorcycles approaching in the near or far distance, 30 were pictures of empty roads and 30 were fillers containing mixed traffic. The fillers were included in the experiment firstly because there is a higher frequency of cars on the road compared to motorcycles and secondly to show a natural range of distances of approaching vehicles. Only the pictures showing vehicles in the mid-distance were of interest to the analysis, as these were expected to result in the most difficult decision-making, with some road users declaring the roadway safe-to-enter and others regarding these distances as unsafe. When a car or motorcycle was present in the nearest lane, it was either at a near, middle or far distance. The near distance was so near that it was clearly unsafe to pull out from the T-junction, and the far distance was safe to pull out – the pictures with far and near vehicles were also fillers. The middle distance was the distance where people were expected to be most hesitant about pulling out – 65 metres away from the T-junction (compare Figures 4.2a, 4.2b and 4.2c) – and so decisions about these scenes were of most interest.

The remaining 40 pictures were taken from the driver's perspective when actually driving (see Figures 4.3a and 4.3b), and were for use in Study 2. The judgement in this experiment concerned the safety of a manoeuvre from the left (nearside) lane to the right (offside) lane, in anticipation of overtaking a slower vehicle, for example. The roads were dual-carriageways and the driver was always in the left lane. The picture showed part of the dashboard, and partial views from the front and side windows, including the driver's door mirror.

In the pictures where the side mirror could be seen, 10 had cars in the mirror, 10 had a PTWV in the mirror, 10 had no vehicle and 10 were fillers to increase the frequency of cars.

Figure 4.2   Photographs showing a car approaching at a) far, b) middle, or c) near distance

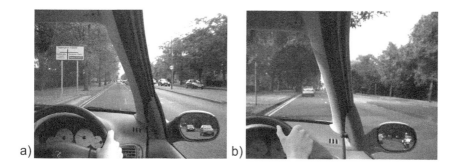

Figure 4.3   Picture from driver's perspective on a dual-carriageway – a) shows a car in the side mirror, and b) shows a PTWV in the side mirror

The vehicles in both the T-junction and the side-mirror stimuli were either of a high or low saliency. Figure 4.4 shows an example from Study 1. The relative saliency of the critical vehicle was estimated by running the picture through a Matlab algorithm based on the Itti and Koch saliency map. Pictures of high saliency vehicles were in the 70th percentile, meaning that 70 per cent of the pixels in the

**Figure 4.4   Picture from driver's perspective on a dual-carriageway –
a) shows a car in the side mirror, and b) shows a PTWV in the
side mirror**

picture were of a lower saliency than the pixels in the area of interest (containing
the vehicle). Low saliency pictures were in the 30th percentile, meaning that only
30 per cent or less of the pixels in the picture were of a lower saliency to those in
the area of interest (containing the vehicle).

An example of a saliency map generated by the Itti and Koch (2000) algorithm
is shown in Figure 4.5. The original photograph is shown in Figure 4.5a, with
the two highest saliency peaks identified by circles – they correspond to the
rider's helmet and the motorcycle fuel tank. The raw saliency map is shown in
Figure 4.5b, with variations in colour, intensity and orientation highlighted by
illuminated regions.

An SR EyeLink II eye tracker was used to record eye movements during the
experiment, sampling at 500 Hz. Pictures were displayed on a 48 cm computer
monitor placed 60 cm from the participants. A nine-point calibration procedure
was used with each participant at the start of the study.

**Figure 4.5   Outputs from the Itti and Koch (2000) saliency algorithm to
identify the two most salient points (left), and showing the full
raw saliency map (right)**

Procedure

Participants were shown on-screen instructions and were also talked through the procedure and given some practice pictures to get used to the button presses. Participants were first shown T-junction pictures one at a time. They had to decide whether they thought it was safe or unsafe to pull into the nearest lane. Traffic could be approaching from the left or the right, but either way the task was always to pull into the nearest lane, in the direction the traffic was flowing.

Participants were told that all roads had 40 mph restrictions (approx. 65 kph), and that all the traffic was travelling at this same speed. Each picture appeared for as long as the participant needed to make a button press response. They pressed 'safe' if they thought they would pull out into the road and 'unsafe' if they thought they would not pull out. Participants were told that there were no correct or incorrect responses, and that it was their opinion that should guide their response. After each picture, a 'drift-correct' fixation spot appeared in order for the eye tracker to be recalibrated, in case the participant's head had moved during the trial.

After inspection of 120 T-junction pictures in Study 1, the participants took a short break and were given a second set of instructions. For Study 2 they were told that the pictures were going to change to the perspective of a driver in the left-hand lane of a dual-carriageway. It was explained that the task now was to decide whether, for each picture, it was safe or unsafe to pull out into the right-hand lane, as they might do in order to overtake a slower vehicle, or in preparation for making a turn at a junction. Participants pressed the 'safe' button if they would pull into the right lane or the 'unsafe' button if they would not. The pictures appeared for as long as was needed to make a response, and were followed by a drift-correct fixation point. In both studies initial fixation was in the centre of the screen.

Results

Results from the two studies will be presented separately, and behavioural measures (percentage of 'safe to manoeuvre' responses, and time taken to execute that decision) and eye movement measures will also be presented separately. Further details of the results from Study 1 are available in Underwood et al. (2011).

Behavioural Measures – Deciding about Safety in Study 1 (Junction Decision)

Participants decided whether or not it would be safe to join the main roadway, in pictures showing a T-junction, and their decisions and response times were recorded. In all of the following analyses we consider only responses to pictures in which a vehicle is shown – the filler trials were discarded. The behavioural and eye-tracking measures are shown in Figures 4.6–4.18.

Differences in safe/unsafe responses (see Figure 4.6) were inspected with a single ANOVA, and the three factors were road user (driver/rider), approaching vehicle (car/motorcycle) and visual saliency of the approaching vehicle (high/

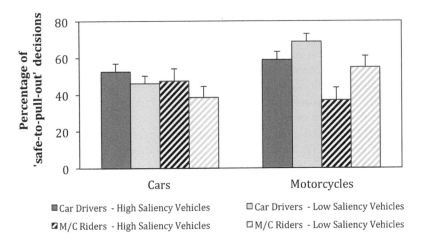

Figure 4.6   The percentages of decisions that it would be safe to pull out into the junction, in the presence of a car or a motorcycle, in Study 1

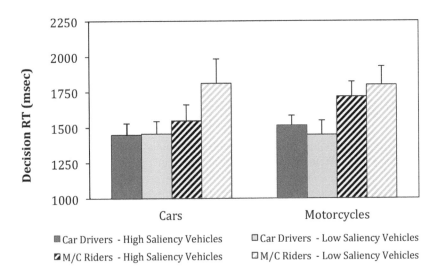

Figure 4.7   The time taken to decide that it would be safe to pull out into the junction in the presence of a car or a motorcycle, in Study 1

low saliency). Only the main effect of vehicle type was reliable, with more 'safe' decisions when the nearest vehicle was a car than a motorcycle, $F(1,75)=11.39$, MSE=0.541, $p < .01$. A two-way interaction between road users and vehicle type, $F(1,75)=5.00$, MSE=0.237, $p < .05$, inspected with an analysis of simple main effects, indicated that drivers responded 'safe' more when a motorcycle was shown than when a car was shown, $F(1,150)=22.45$, MSE=1.066, $p < .001$, and that riders responded similarly to both types of vehicle. A second two-way interaction, between vehicle and saliency, $F(1,75)=25.90$, MSE=0.813, $p < .001$, indicated that only decisions about high saliency vehicles were similar, but that low saliency motorcycles elicited more 'safe' decisions than did low saliency cars, $F(1,150)=41.11$, MSE=1.621, $p < .001$.

Decision times (see Figure 4.7) were also analysed with a three-factor ANOVA. Only the two-way interaction involving road users and visual saliency was reliable, $F(1,75)=6.91$, MSE=0.712, $p < .01$. Simple main effects indicated that whereas riders responded more slowly to low saliency than to high saliency vehicles, $F(1,75)=7.98$, MSE=0.822, $p < .01$, there was no difference in the response times of drivers.

Eye Movement Measures of Picture Inspection in Study 1 (Junction Decision)

Eye movement data were analysed to establish the attention given to the critical vehicle itself, looking at the number of fixations on the vehicle, and the duration of the first fixation on the vehicle, as well as eye movements resulting in fixation on the vehicle. To determine fixations on the vehicle itself, we first created an 'area of interest' (AoI) around the largest vehicle to be considered from all photographs used in the experiment, and then centred this AoI on the vehicle, however small, in each picture. PTWVs invariably occupied less space in the photographs, as they do on actual roadways. By using a constantly sized AoI we ensured that any variations in fixation behaviour were not associated with differently sized AoIs. Data from one motorcyclist were lost from these analyses, due to poor eye-tracking calibration. Examples of the fixation plots for drivers and riders inspecting a high and a low saliency motorcycle are shown in Figure 4.8.

The number of fixations made prior to fixation on the vehicle (see Figure 4.9) was inspected with a three-factor ANOVA that indicated main effects of the vehicle, $F(1,74)=33.02$, MSE=3.104, $p < .001$ (more fixations prior to inspection of a car than a motorcycle), and of saliency, $F(1,74)=22.15$, MSE=1.56, $p < .001$ (fewer fixations prior to inspection of high saliency vehicles). Vehicle type also interacted with vehicle saliency, $F(1,74)=37.52$, MSE=2.70, $p < .001$, and an analysis of simple main effects showed a difference between vehicles only in the low saliency conditions, $F(1,148)=77.16$, MSE=6.404, $p < .001$, when there were fewer fixations prior to inspection of motorcycles.

High saliency motorcycle

Low saliency motorcycle

Figure 4.8 **Fixations made by all car drivers (upper and lower left pictures) and all motorcycle riders (upper and lower right pictures) when inspecting an approaching motorcycle that was estimated as being of high saliency (top pictures) or low saliency (bottom pictures)**

The duration of the fixation immediately preceding inspection of the vehicle (see Figure 4.10) was analysed with an ANOVA that indicated no main effects, and one interaction. Vehicle-type interacted with saliency, $F(1,74)=4.11$, MSE=0.066, $p < .05$, with simple main effects indicating shorter fixations for motorcycles with low relative to high saliency, $F(1,148)=7.41$, MSE=0.108, $p < .01$.

The number of fixations on the vehicle (see Figure 4.11) was inspected with an ANOVA that indicated main effects of the vehicle, $F(1,74)=12.59$, MSE=7.32, $p < .001$ (more fixations on motorcycles than cars). Vehicle-type also interacted with saliency, $F(1,74)=11.22$, MSE=4.37, $p < .01$, with simple main effects indicating a difference between vehicles only in the low saliency conditions, $F(1,148)=28.19$, MSE=13.680, $p < .001$, when there were more fixations on low saliency motorcycles.

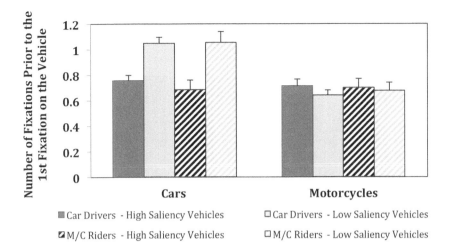

Figure 4.9 **The number of fixations made on the roadway scene prior to fixating the approaching car or motorcycle, in Study 1**

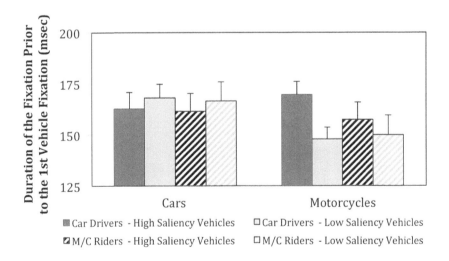

Figure 4.10 **The duration of the fixation immediately preceding the first fixation on the approaching car or motorcycle, in Study 1**

The duration of the first fixation on the vehicle (see Figure 4.12) was also analysed with an ANOVA that indicated a main effect of the vehicle-type, $F(1,74)=5.66$, $MSE=0.065$, $p < .05$ (longer first fixations on cars than on motorcycles). There were no interactions.

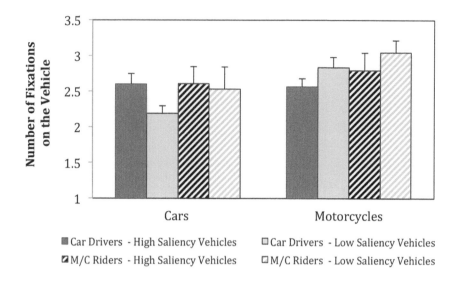

Figure 4.11 The total number of fixations made on the approaching car or motorcycle, in Study 1

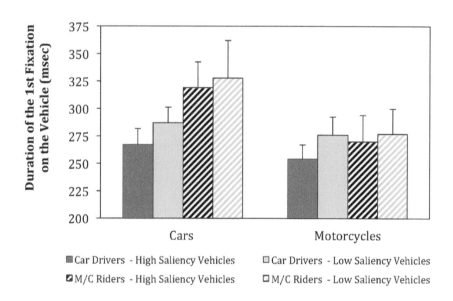

Figure 4.12 The duration of the first fixation on the approaching car or motorcycle, in Study 1

Behavioural Measures – Deciding about Safety in Study 2 (Changing Lane)

In the second study participants decided whether or not it would be safe to move from the nearside lane to the outside lane of a dual-carriageway, in pictures showing the view ahead of a driver that included the external door mirror (see Figure 4.3). As in the first study, eye movements were recorded as well as their decisions and response times. The behavioural and eye-tracking measures from this study are shown in Figures 4.13–18.

Differences in safe/unsafe responses (see Figure 4.13) were inspected with a single ANOVA, with the three factors of road user (driver/rider), approaching vehicle (car/motorcycle) and visual saliency of the approaching vehicle (high/low saliency). The type of vehicle in the mirror influenced the decision made about the manoeuvre, $F(1,75)=16.84$, MSE=0.754, $p < .001$, with more 'safe to change lane' responses when a car rather than a motorcycle was shown. There was also a main effect of visual saliency, with more declarations of 'safe' when a low saliency vehicle was shown. There were no other main effects or interactions.

Decision times (see Figure 4.14) were also analysed with a three-factor ANOVA that showed just one effect, that of visual saliency, $F(1,75)=10.62$, MSE=4.45, $p < .01$. Faster responses were made to high saliency vehicles. There were no other main effects or interactions.

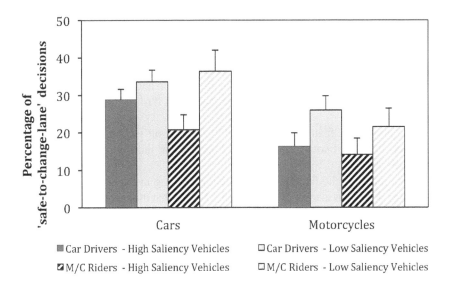

Figure 4.13 **The percentage of decision that it would be safe to change lane, in the presence of an image of a car or a motorcycle in the door mirror, in Study 2**

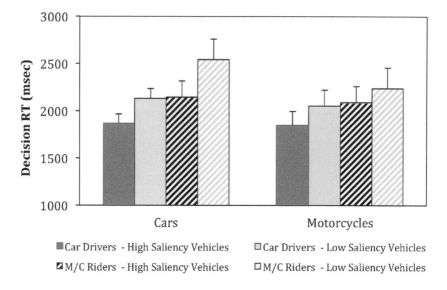

Figure 4.14   The time taken to make the decision that it would be safe to change lanes, in Study 2

Eye Movement Measures of Picture Inspection in Study 2 (Changing Lane)

The measures of visual attention used in Study 1 were again recorded here, to establish the number of eye fixations on the critical vehicle shown in the mirror, fixations before looking at the vehicle in the mirror and the fixations on that vehicle. Data from two of the PTWV riders were lost through poor calibration of the eye-tracker.

The numbers of fixations made prior to fixation on the vehicle (see Figure 4.15) were inspected with a three-factor ANOVA that indicated no main effects and no interactions.

The duration of the fixation immediately preceding inspection of the vehicle (see Figure 4.16) was also analysed with an ANOVA that indicated no main effects and no interactions.

The numbers of fixations on the vehicle in the mirror (see Figure 4.17) were inspected with a three-factor ANOVA that showed a main effect of saliency, $F(1,73)=10.42$, MSE=7.66, $p < .01$, in which low saliency vehicles received more fixations prior to the decision being indicated. There were no other main effects and no interactions.

The duration of the first fixation on the vehicle in the mirror (see Figure 4.18) was also analysed with a three-factor ANOVA. There were no main effects, and the only interaction involved road users and visual saliency, $F(1,73)=6.12$, MSE=41731, $p < .05$, and an analysis of simple main effects indicated that riders

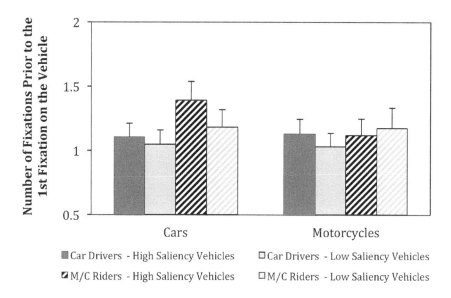

Figure 4.15   The number of fixations made on the roadway scene prior to fixating the approaching car or motorcycle, in Study 2

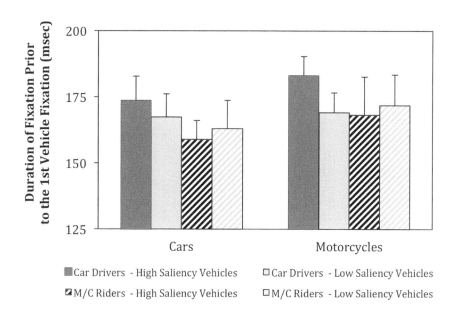

Figure 4.16   The duration of the fixation immediately preceding the first fixation on the approaching car or motorcycle, in Study 2

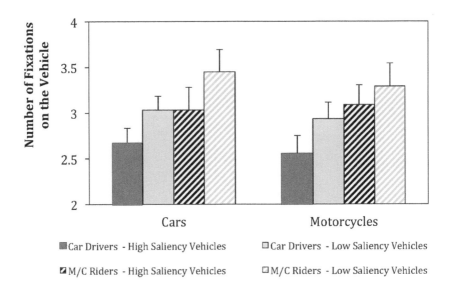

Figure 4.17   The total number of fixations made on the approaching car or motorcycle, in Study 2

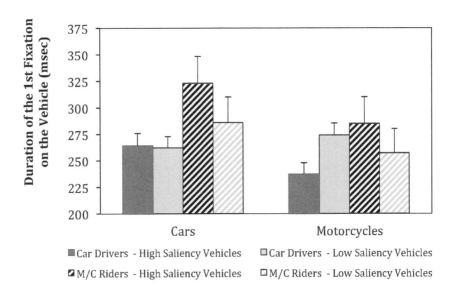

Figure 4.18   The duration of the first fixation on the approaching car or motorcycle, in Study 2

and drivers had similar fixation durations when looking at low saliency vehicles, but that riders had longer fixations than drivers when looking at high saliency vehicles, F(1,146)=8.26, MSE=95658, p < .01. When asked informally about these long fixations, the motorcycle riders gave an interested though admittedly anecdotal explanation. When they saw a motorcycle, they sometimes suggested, they assessed the power of the vehicle before deciding whether to make the manoeuvre. A low-powered motorcycle would allow a manoeuvre that would be precluded by the presence of a more powerful vehicle, and that power could be best assessed by a long inspection.

Discussion

In the first experiment drivers and riders made decisions about whether they would enter a roadway at a T-junction. Sometimes a vehicle was present – a car or a PTWV – at a distance sufficient for the participant to have to think carefully before responding. The behavioural and eye movement measures indicated effects with all analyses, and these will be first summarised here, starting with the effects of visual saliency as this was the main interest of the experiment.

Saliency had several effects in Study 1: all road users were more likely to regard the scene as 'safe' when a motorcycle had low saliency, and riders were especially slower at making decisions about low saliency vehicles. Changing the appearance of a PTWV had an effect of encouraging more cautious responses with higher saliency motorcycles (that is, fewer 'safe to pull out' responses). The ability of a vehicle to attract attention (as opposed to its ability to hold attention once gained) is indicated by variations in fixation behaviour prior to the vehicle's inspection. Pre-inspection processing of the vehicle was assessed here by recording the number of eye fixations prior to inspection, the duration of the fixation immediately preceding inspection and the length of the saccadic movement that resulted in fixation on the vehicle. Visual saliency was potent prior to inspection of the vehicle in the roadway, with longer saccadic movements towards high saliency vehicles (both cars and PTWVs). These longer saccades were associated with fewer fixations prior to inspection of a high-saliency vehicle, but this held especially for cars: low saliency cars attracted attention particularly slowly according to this measure. Pre-inspection fixation durations tended to be shorter for motorcycles with low saliency. Inspection of the vehicle itself also showed an effect of visual saliency, with more fixations on low saliency motorcycles prior to execution of the safe/ unsafe decision.

There were few differences between drivers and riders, and the most striking of these was the tendency of drivers (but not riders) to declare as 'safe' roadways in which a motorcycle was present. Riders also gave faster decisions about high rather than low saliency vehicles, whereas drivers (who tended to respond slightly faster than riders) did not distinguish between high and low saliency vehicles.

Decisions declaring the situation as being 'safe to pull out' were more frequent when a motorcycle was the vehicle in the roadway (as it was in the video-based study reported by Crundall et al., 2012), although this trend was moderated when it was a high saliency motorcycle. There was no difference in judgements when high saliency vehicles were shown, but low saliency motorcycles attracted more decisions to enter the junction. There were more eye fixations on motorcycles than on cars, immediately prior to a decision, and the first fixation was shorter on motorcycles than on cars. In the Crundall et al. study drivers made longer fixations on cars while riders made longer fixations on PTWVs, but this interaction was not apparent in the present experiment.

The measures in Study 2, in which the decision involved whether it would be safe to move into the outer lane of a two-lane highway, again showed extensive effects of the saliency of the vehicle shown, even though the location of the vehicle in the photograph was more predictable than it was in Study 1. If there was a low saliency vehicle shown in the mirror, road users were more likely to say that they could change lane – into the path of that vehicle – than if it was a high saliency vehicle present. Decisions were also faster when it was a high saliency vehicle visible in the mirror, and this is associated with eye fixation behaviour. There were fewer fixations on high saliency vehicles. In addition, motorcycle riders had longer first fixations on high saliency vehicles.

There is a long-standing belief amongst accident researchers that the use of conspicuity enhancements through daytime-running lights or high visibility clothing can minimize the crash liability of road users, with data from the analysis of crash reports providing support (Thomson, 1980; Zador, 1985; Cercarelli et al., 1992; Farmer and Williams, 1992; Elvik, 1993; Radin et al., 1996; Yuan, 2000). On-road trials suggest the same conclusion (Olson et al., 1979, 1981; Wood et al., 2012). The present studies, and others reported in this volume, provide evidence from laboratory studies than enable precise control over presentations and testing conditions that a conspicuous motorcycle is more likely to be dealt with more cautiously by car drivers.

The two studies set out to determine the effects of visual saliency on decisions about roadway manoeuvres. Does it make a difference to the decision about pulling out of a junction, or changing lane for an overtaking manoeuvre if a vehicle in the roadway is more conspicuous? Whereas it might seem obvious that a high saliency vehicle is more likely to be seen, a dispute in the scene perception literature casts doubt on the efficacy of saliency to attract attention. Secondary questions involved differences between car drivers and motorcycle riders making the decisions, and differences between decisions when a car or a motorcycle was present.

When a vehicle in the scene was more conspicuous, by virtue of being more colourful or brighter than its background, it elicited more cautious decisions (more 'unsafe' responses) that were also made more quickly than when the vehicle had low saliency. This pattern of faster, more cautious responding was seen in both studies: when high saliency vehicles were present then decisions about entering a junction and decisions about moving into an outer lane on a two-lane highway were made more quickly and with greater caution. Conspicuous vehicles elicit more cautious decision-making in other road users.

Eye movement analyses indicated that the slower decisions for low saliency vehicles were associated with a greater number of fixations on the vehicle itself, and a longer first fixation. Low saliency induced greater inspection of the vehicle, and again this pattern was observed with both the junction and the lane-change decisions. The influence of saliency was seen prior to inspection of the vehicle, and this lends support to the Itti and Koch (2000) model of attention, whereby eye fixations are initially guided by a low-level visual saliency map.

Three caveats should be taken into consideration when interpreting these results. Firstly, the studies used still photographs of roadway scenes, presented as part of a laboratory experiment. There is a question here about the applicability of laboratory results to actual real-world roadway behaviour: would road users behave in the same way when driving/riding, or is their laboratory behaviour determined by the demand-characteristics of the laboratory? Two recent laboratory experiments have used roadway video recordings to investigate the decision-making and allocation of visual attention of drivers and riders. Crundall et al. (2012) and Shahar et al. (2012) used the same T-junction and door mirror inspection tasks as here, asking whether it is safe to pull out and whether it is same to change lane. Drivers and riders watched video recordings filmed from a car manoeuvering on actual roads, while their decisions and eye movements were recorded. Some of the results matched those from the experiments here – riders were faster at making 'safe to act' decisions than drivers, both studies suggest that riders had longer fixations on PTWVs than did drivers, and there was more cautious responding in the presence of a PTWV (especially by riders). This suggests that we may be able to generalize beyond the use of still photographs to moving images of roadway scenes. Furthermore, the results from the Olson et al.'s (1979, 1981) quasi-experimental study in which the behaviour of actual road users was observed in the presence of conspicuous or less conspicuous PTWVs suggest that the results from these laboratory experiments can be generalized to behaviour by road users in natural situations. The second caveat is that the calculations of conspicuity in photographs are made on the basis of regional image change in the photograph. Conspicuity is determined relative to the one background in the photograph. Accordingly, a region or an object identified as highly conspicuous would not necessarily be conspicuous when presented against a different background, as would be the case of a vehicle travelling along different roads and in different lighting conditions. Finally, we have no way of assessing the risk compensation mechanisms that would operate if PTWV riders were required to make themselves more conspicuous. Would they regard themselves as then being more visible and accordingly more safe, and then make more risky roadway decisions themselves? Underwood et al. (1993) point to a number of cases where the introduction of safety legislation such as the compulsory use of helmets by PTWV riders resulted in an overall increase in the number of crashes. Naturalistic observation could address this question by assessing the current behaviour of conspicuous and less conspicuous PTWV riders in terms of risks taken at junctions, and when overtaking and so on.

References

Cercarelli, L.R., Arnold, P.K., Rosman, D.L., Sleet, D. and Thornett, M.L. (1992). Travel exposure and choice of comparison crashes for examining motorcycle conspicuity by analysis of crash data. *Accident Analysis and Prevention*, 24, 363–8.

Crundall, D., Crundall., E., Clarke, D. and Shahar, A. (2012). Why do car drivers fail to give way to motorcycles at t-junctions? *Accident Analysis and Prevention*, 44, 88–96.

Elvik, R. (1993). The effects on accidents of compulsory use of daytime running lights for cars in Norway. *Accident Analysis and Prevention*, 25, 383–98.

Farmer, C.M. Williams, A.F. (2002). Effects of daytime running lights on multiple-vehicle daylight crashes in the United States. *Accident Analysis and Prevention*, 34, 197–203.

Findlay, J.M. and Walker, R. (1999). A model of saccade generation base on parallel processing and competitive inhibition. *Behavioral and Brain Sciences*, 4, 661–721.

Foulsham, T. and Underwood, G. (2007). How does the purpose of inspection influence the potency of visual saliency in scene perception? *Perception*, 36, 1123–38.

Fuchs, I., Ansorge, U., Redies, C. and Leder, H. (2011). Salience in paintings: Bottom-up influences on eye fixations. *Cognitive Computation*, 3, 25–36.

Henderson, J.M., Brockmole, J.R., Castelhano, M.S. and Mack, M. (2007). Visual saliency does not account for eye movements during visual search in real-world scenes. In van Gompel, R.P.G., l, Fischer, M.H., Murray, W.S. and Hill, R.L. (eds), *Eye Movements: A Window on Mind and Brain*, 537–62. Elsevier, Oxford.

Henderson, J.M., Weeks, P.A. and Hollingworth, A. (1999). The effects of semantic consistency on eye movements during complex scene viewing. *Journal of Experimental Psychology: Human Perception and Performance*, 25, 210–28.

Humphrey, K. and Underwood, G. (2010). The potency of people in pictures: Evidence from sequences of eye fixations. *Journal of Vision*, 10(10): 19, 1–10.

Itti, L. and Koch, C. (2000). A saliency-based search mechanism for overt and covert shifts of visual attention. *Vision Research*, 40, 1489–506.

Lamy, D., Leber, A. and Egeth, H.E. (2004). Effects of task relevance and stimulus-driven salience in feature-search mode. *Journal of Experimental Psychology: Human Perception and Performance*, 30, 1019–31.

Navalpakkam, V. and Itti, L. (2005). Modeling the influence of task on attention. *Vision Research*, 45, 205–31.

Nothdurft, H.-C. (2002). Attention shifts to salient targets. *Vision Research*, 42, 1287–306.

Olson, P.L., Halstead-Nussloch, R. and Sivak, M. (1979). Development and testing of techniques for increasing the conspicuity of motorcycles and motorcycle drivers. *Final Report DOT-HS-6–01459, National Highway Traffic Safety Administration.* Washington, DC.

Olson, P.L., Halstead-Nussloch, R. and Sivak, M. (1981). The effect of improvements in motorcycle/motorcyclist conspicuity on driver behavior. *Human Factors*, 23, 237–48.

Parkhurst, D., Law, K. and Niebur, E. (2002). Modelling the role of salience in the allocation of overt visual attention. *Vision Research*, 42, 107–23.

Radin, U.R.S., Mackay, G.M. and Hills, B.L. (1996). Modelling of conspicuity related motorcycle accidents in Seremban and Shah Alam, Malaysia. *Accident Analysis and Prevention*, 28, 325–32.

Shahar, A., van Loon, E., Clarke, D. and Crundall, D. (2012). Attending overtaking cars and motorcycles through the mirrors before changing lanes. *Accident Analysis and Prevention*, 44, 104–10.

Stirk, J.A. and Underwood, G. (2007). Low-level visual saliency does not predict change detection in natural scenes. *Journal of Vision*, 7(10): 3, 1–10,

Tatler, B.W. (2007). The central fixation bias in scene viewing: Selecting an optimal viewing position independently of motor biases and image feature distributions. *Journal of Vision*, 7(14): 4, 1–17.

Thomson, G.A. (1980). The role frontal motorcycle conspicuity has in road accidents. *Accident Analysis and Prevention*, 12, 165–78.

Torralba, A., Oliva, A., Castelhano, M. S. and Henderson, J.M. (2006). Contextual guidance of eye movements and attention in real-world scenes: The role of global features in object search. *Psychological Review*, 113(4), 766–86.

Underwood, G. and Foulsham, T., (2006). Visual saliency and semantic incongruency influence eye movements when inspecting pictures. *Quarterly Journal of Experimental Psychology*, 59, 1931–49.

Underwood, G., Foulsham, T. and Humphrey, K. (2009). Saliency and scan patterns in the inspection of real-world scenes: Eye movements during encoding and recognition. *Visual Cognition*, 17, 812–34.

Underwood, G., Foulsham, T., van Loon, E., Humphreys, L. and Bloyce, J. (2006). Eye movements during scene inspection: A test of the saliency map hypothesis. *European Journal of Cognitive Psychology*, 18, 321–42.

Underwood, G., Humphrey, K. and van Loon, E. (2011). Decision about objects in real-world scenes are influenced by visual saliency before and during their inspection. *Vision Research*, 51, 2031–8.

Underwood, G., Jiang, C. and Howarth, C.I. (1993). Modelling of safety measure effects and risk compensation. *Accident Analysis and Prevention*, 25, 277–88.

Underwood, G., Jebbett, L. and Roberts, K. (2004). Inspecting pictures for information to verify a sentence: Eye movements in general encoding and in focused search. *Quarterly Journal of Experimental Psychology*, 57A, 165–82.

Underwood, G., Templeman, E., Lamming, L. and Foulsham, T. (2008). Is attention necessary for object identification? Evidence from eye movements during the inspection of real-word scenes. *Consciousness & Cognition*, 17, 159–70.

Wood, J.M., Tyrrell, R.A., Marszalek, R., Lacherez., P. and Chu, B.S. (2012). Using reflective clothing to enhance the conspicuity of bicyclists at night. *Accident Analysis and Prevention*, 45, 726–30.

Yuan, W. (2000). The effectiveness of the 'ride-bright' legislation for motorcycles in Singapore. *Accident Analysis and Prevention*, 32, 559–63.

Zador, P.L. (1985). Motorcycle headlight-use laws and fatal motorcycle crashes in the US, 1975–83. *American Journal of Public Health*, 75, 543–6.

Chapter 5

'Should I Stay or Should I Go?' Examining the Effect of Various Conspicuity Treatments on Drivers' Turning Performance

Eve Mitsopoulos-Rubens and Michael G. Lenné

The Problem

To say that powered two-wheeler (PTW) riders are among the most vulnerable of road users is by no means an overstatement. PTW riders' vulnerability is reflected in the over-representation of PTWs in the crash statistics. Drawing on recent data from the Australian state of Victoria (Transport Accident Commission, 2013), 15 per cent of road users who were killed in 2012 were PTW riders and pillion passengers. However, during the same period, PTWs accounted for only about 4 per cent of the number of vehicles registered in Victoria. In Australia, as well as in other motorised countries worldwide, PTW riding continues to gain in popularity. It is perhaps not surprising, therefore, that understanding the factors which contribute to PTW crash risk and implementing measures to reduce that risk is of interest to both road safety researchers and practitioners alike.

The 'Diminished Conspicuity' Hypothesis

A prominent theme to have emerged in PTW safety relates to the concept of conspicuity. Diminished PTW conspicuity has been implicated as a factor in crashes involving multiple vehicles, most notably collisions between a passenger car and a PTW. The significance of the issue is clear when considering the magnitude of PTW crashes involving a passenger vehicle. In the European-based Motorcycle Crash In-Depth Study (MAIDS), it was reported that PTW crashes involving a passenger vehicle accounted for 63.1 per cent of cases in urban areas (ACEM, 2009).

Right-of-way violations are a prominent multi-vehicle crash type of PTW riders. In an in-depth analysis of almost 2,000 PTW crashes that occurred in the UK (spanning the years 1997–2002), Clark et al. (2007) found that 38 per cent of crashes could be characterised as right-of-way collisions, with the majority of these taking place at T-intersections. Moreover, in the overwhelming majority of these cases (about 80 per cent) it was the driver of the passenger vehicle, as opposed to the PTW rider, who was found to be at fault. That is, it was the PTW rider who had right-of-way – as might be the case when the car is turning onto a road across the path of an oncoming PTW. Similar patterns have also been observed recently in the Australian context. In a case-control study of serious injury PTW crashes that is currently underway (Allen et al., 2013), analysis of cases recruited to date (n=75) revealed that 60 per cent of the crashes involved multiple vehicles. Of these multi-vehicle crashes, 67 per cent were reported to occur at an intersection. Of these intersection crashes, the most prevalent type (69 per cent) involved another vehicle turning across the path of a PTW.

It follows that crashes involving another vehicle turning across the path of a PTW are a crash type that could benefit from improvements in PTW conspicuity and, more specifically, improvements in PTW 'sensory' conspicuity. An object's sensory conspicuity relates to how well a given object can be distinguished visually from its background on the basis of the object's physical characteristics (for example, Engel, 1976; Cole and Jenkins, 1984; Wertheim, 2010). Various treatments for augmenting the sensory conspicuity of PTWs have been proposed. These include bright clothing and lighting, most notably in the form of daytime-running lights (DRLs) or low-beam headlights. The key is to optimise the contrast between the PTW (including the PTW rider) and the surrounding environment, and thereby maximise the PTW's detectability (for example, Hole et al., 1996; Rogé et al., 2010). In the context of a vehicle waiting to turn across the path of approaching vehicles, DRLs or low-beam headlights, for example, would work by increasing the chances of other road users detecting sufficiently early the approach of an oncoming PTW.

The assumption is that more efficient detection of a PTW will lead to more efficient and cautious decision-making and, thus, safer driving by other motorists in the vicinity of PTWs. However, much of the PTW sensory conspicuity research to date has focused on the effect of specific treatments on PTW detection rates and perception accuracy. There is thus a need for additional, systematic research to understand the effects of specific PTW sensory conspicuity treatments on drivers' actual decision-making and driving actions.

PTW Low-beam Headlights and Drivers' Turning Performance

In response to this fundamental gap in knowledge, we conducted a driving simulator experiment, which sought to examine the influence of a sensory conspicuity treatment on drivers' decisions to turn across the path of an oncoming PTW in daylight conditions. We chose to focus on low-beam headlights as the sensory conspicuity treatment of interest. The main question being asked in this experiment was:

> Is an oncoming vehicle that is a PTW with low-beam headlights on associated with a lower rate of gap acceptance, a more efficient turn

strategy and larger safety margins when turning than an oncoming vehicle that is a PTW with low-beam headlights off?

The method and findings of this experiment are described below. This research was carried out as part of Work Package 5.2 of the European Commission 7th Framework Project – '2-Be-Safe'. A subset of the results presented here have been published elsewhere (Mitsopoulos-Rubens and Lenné, 2012).

What Did We Do?

Driving Simulator Trials

The study involved the use of the 'Portable Simulator Facility' at the Monash University Accident Research Centre (MUARC) in Melbourne, Australia. The simulator itself is an Eca Faros EF-X and consists of a small cab with vehicle parts, including an adjustable seat, pedals, steering wheel, gear box and seat belt. The visual images for the driving scenarios are presented on three 19-inch LCD monitors, providing a field of view of 120 degrees. To add to the driver's experience, stereophonic sound (that is, engine noise and the sound of other traffic) is delivered through speakers.

A gap-acceptance task characterised each of the simulator trials. This task was modelled on that developed previously by Eve Mitsopoulos-Rubens as part of her PhD research (Mitsopoulos-Rubens, 2010), where the task had been used successfully to differentiate the gap acceptance and turning performance of novice and experienced drivers.

Each of the simulator trials contained a single right-turn event. The trials utilized a two-lane urban road (speed limit of 60 km/h) with a single unsignalized 'T' intersection. Weather conditions were clear and all trials simulated daytime lighting.

All simulator trials had the same basic design (see Figure 5.1). As the participant in the simulator cab ('own-cab') travelled towards the intersection, a series of four vehicles (all travelling at 60 km/h) approached from the opposite direction in the adjacent lane. Participants were asked to drive at a speed of 60 km/h (that is, the speed limit) unless the task required otherwise (that is, slowing down at the T-intersection to turn) and to turn right at the intersection. With regards to the timing of the turn, participants were told that they should turn at the first opportunity where they felt that they had enough room in the stream of oncoming traffic to turn through safely. At the intersection, the positioning of the oncoming vehicles relative to own-cab was such that participants had only two options: either to attempt to turn through the gap between the second vehicle and the third vehicle ('target vehicle') or to wait until all four vehicles had passed the intersection before turning.

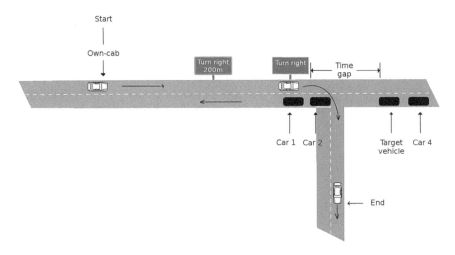

Figure 5.1   High-level schematic of the simulator trials

Across trials, the 'target vehicle' was of two main types: (1) a PTW with headlights on and (2) a PTW with headlights off. When in the 'on' position, the PTW headlights were programmed to resemble, to the extent possible, the colour and brightness of low-beam headlights on a motorcycle. Critically, an analysis of the PTWs' saliency showed that when the PTW headlights were on, the PTW was the most salient object in the image. In contrast, when the PTW headlights were off, the PTW did not feature in the top six objects identified as the most salient objects in the image. This analysis was carried out for us by Geoff Underwood at the University of Nottingham using the Itti and Koch (2000) saliency algorithm.

Other than when the target vehicle was a PTW, all vehicles in the scenario were medium-sized cars. To reduce target vehicle predictability across the simulator trials, a passenger car with headlights off was used as the target vehicle in a subset of the trials.

The variable 'time gap' was also manipulated across trials. Time gap is the size of the gap (in seconds) between oncoming vehicles that participants must choose as either acceptable or unacceptable in deciding when to turn. Time gap has been established previously as a primary determinant of one's gap-acceptance decisions (for example, Caird and Hancock, 2002) and so was felt necessary for consideration here. Time gap was operationalized in the current experiment as the distance in seconds between the front of the second oncoming vehicle (car 2 in Figure 5.1) and the front of the target vehicle.

For each target vehicle, we wanted to expose participants to three time gaps: one short, one medium and one long. The intention was that the short and long time gaps represent gaps with low and high probability of gap acceptance,

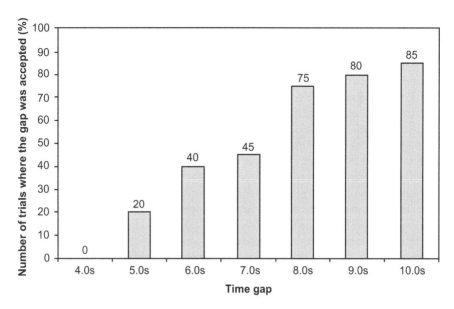

Figure 5.2   Proportion of pilot study trials for each time gap where the gap was accepted

respectively. The medium gap is intended to represent time gaps in the decision dilemma zone, with approximately 50 per cent likelihood of being accepted. Selecting the actual time value for each level of the time gap variable involves some experimentation itself to ensure that the values chosen are appropriate for the particular simulated driving environment. For this reason, we conducted a pilot study to establish which three time gap values we should use in the experiment proper. Pilot study participants were 10 fully licensed car drivers with a mean age of 37.8 years (SD=11.4) and who had been licensed to drive a car for an average of 15.5 years (SD=6.9).

Seven time gaps were tested in the pilot study: 4.0, 5.0, 6.0, 7.0, 8.0, 9.0 and 10.0 seconds. Each time gap was represented in a single trial. Across trials, the target vehicle was always a passenger car, and all vehicles were coloured dark green-blue and had their headlights off. Participants completed each trial twice, with the 14 trials presented to participants in one of three pseudo-random orders. Participants first received four practice trials. Across the seven time gaps, the proportion of trials where the gap was accepted ranged from none (4.0 seconds) to 85 per cent (10.0 seconds). This is reflected in Figure 5.2. The three time gaps eventually selected for the final study – 5.0, 7.0 and 9.0 seconds – were those associated in the pilot study with a 20 per cent (that is, low), 50 per cent (that is, medium) and 80 per cent (that is, high) rate of gap acceptance, respectively.

Participants and Procedure

Forty-three participants (36 males and 7 females) each attended a single study session of approximately one hour at MUARC. All participants held a full car driver's licence and were experienced car drivers, having held their car driver's licence for an average of 17.8 years (range: 4–37 years). The average age of participants was 37.1 years (range: 24–55 years).

Following a brief introduction to the study and completion of a short background questionnaire, participants completed several simulator trials for familiarization and practice. The experimental trials followed. Nine trials were developed using the combinations of the three time gap levels (short, medium, long) and the three target vehicle types (passenger car, PTW headlights off, PTW headlights on). These nine trials were presented twice, along with four trials in which there was no target or fourth oncoming vehicle. These four trials served as 'dummy trials'. All 22 trials were presented to participants in one of four pseudo-random orders. At the conclusion of the session, each participant was compensated a small fee for his/her involvement.

Analysis

Turning performance was defined in terms of several dependent variables: gap acceptance (accept or reject), time taken to travel through the turn and safety margin. Each dependent variable was analysed using a generalized estimating equation (GEE) (Liang and Zeger, 1986). In every case, the structure of the inter-correlation matrix was specified as exchangeable. Further, in the case of gap acceptance, the GEE was specified with binomial distribution and logit link function. For turn travel time and safety margin, the GEE was specified with normal distribution and identify link function. Statistical significance was defined as $p \leq 0.05$.

What Did We Find?

Gap Acceptance

To explore gap acceptance, data were taken from all trials where a target vehicle was present – in particular, those where the target vehicle was a PTW with lights on, and those where it was a PTW with lights off. The aim of the analysis was to explore the influence of target vehicle characteristics on drivers' turning gap acceptance, and whether this effect is contingent on time gap. If sensory conspicuity is an important factor in drivers' decisions as to whether or not to accept a gap when turning across the path of oncoming traffic, then an expectation is that heightened sensory conspicuity (that is, headlights on) would be associated with fewer gap acceptances than relatively lower sensory conspicuity (that is, headlights off), and particularly when crash likelihood is higher (that is, short and medium time gaps).

The analysis revealed a significant time gap main effect ($\chi^2(2)=105.62$, $p<0.01$), and a significant time gap x target vehicle interaction ($\chi^2(4)=10.07$, $p<0.05$). Both the time gap main effect and time gap x target vehicle interaction are reflected in Figure 5.3. With regards to time gap, as intended, participants accepted more gaps with each increment in time gap. In other words, the highest rate of gap acceptance occurred at the longest time gap and the lowest rate was associated with the shortest time gap. As can be seen in Figure 5.3, the interaction appears to have been driven by different rates of gap acceptance across the two target vehicle types at different time gaps. At the short time gap, there were fewer trials where the gap was accepted when the target vehicle was the PTW with lights on than when it was the PTW with lights off. At the medium time gap, the pattern is reversed with more gaps accepted ahead of a PTW with lights on than ahead of the PTW with lights off. This latter observation is quite unexpected as it suggests that participants are more accepting of risk, and, hence, less cautious, under conditions of heightened conspicuity.

To further explore the time gap x target vehicle interaction, an individual GEE was carried out for each time gap. A significant target vehicle main effect was found, but at the short time gap only ($\chi^2(2)=6.21$, $p<0.05$). Examination of the parameter estimates confirms that, at the short gap, the odds of accepting a gap

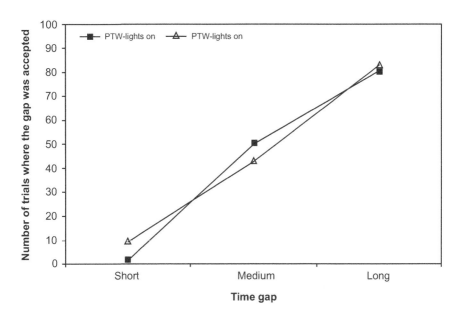

Figure 5.3   **Number of trials for each time gap where the gap was accepted for PTW lights on and PTW lights off**

ahead of a PTW with headlights off were significantly greater than the odds of accepting a gap ahead of a PTW with headlights on (Relative OR=0.127; Relative OR 95 per cent CI=0.024–0.679). The analysis carried out for the medium time did not show that the observed difference between the headlights on and headlights off conditions was a statistically significant one. There is thus insufficient evidence to support 'the less risk-averse hypothesis' posited above with regards to the medium time gap.

Turn Time

The data on turn time were taken from those trials where the target vehicle was a PTW, either with headlights on or off. For all gap-acceptance cases, the turn was divided into three phases: approach, intermediate and completion. Each phase was defined with reference to the point along own-cab's path when the front of own-cab and the front of car 2 were at their closest while still occupying adjacent lanes (see Figure 5.1). The approach phase constituted the 10-metre distance in the range 15 to 25 metres before this point. The intermediate phase represented the 15-metre distance immediately preceding the reference point. The completion phase spanned the distance from the reference point to the entrance of the intersecting road.

In exploring turn time, of interest was whether drivers adopted any regulatory strategies at different points during the turn that involved travelling faster when the oncoming vehicle was a PTW with lights on than when it was a PTW with lights off. It had been established previously that partitioning the turn into the three phases as described above provided important information on drivers' turn strategies (Mitsopoulos-Rubens, 2010). In particular, it was described that less time to travel through the approach phase was associated with early, more proactive decision-making. In contrast, a more efficient completion phase was said to be associated with more reactive, last-chance decision-making.

For the approach phase, the analysis revealed a significant gap acceptance main effect ($\chi^2(1)$=38.25, p<0.01) and a significant time gap x gap-acceptance interaction ($\chi^2(4)$=20.42, p<0.01). This interaction is explained in Table 5.1. At each of the three time gaps, participants took significantly more time to travel through the approach phase when the decision was made to reject the gap than when the decision was made to accept it. This pattern in the data suggests that, as suspected, participants had decided by the approach phase whether or not they were going to accept the gap with which they were presented. Further, participants took significantly less time to travel through the approach phase when the short gap was accepted than when the long gap was accepted. This pattern suggests that, if the decision had been made to accept the gap, participants adapted for the shortest time gap by travelling through the approach phase in less time than they did for the longest time gap.

Table 5.1   **Time gap x gap-acceptance interaction parameter estimates for the time taken to travel through the approach phase**

Parameter	B	SE	Wald chi-square	95% Wald confidence intervals
Short x Accept	0.90	0.04	450.70	0.817–0.983
Short x Reject	1.52	0.09	320.94	1.357–1.691
Medium x Accept	1.04	0.04	688.24	0.965–1.121
Medium x Reject	1.31	0.06	433.34	1.190–1.438
Long x Accept	1.04	0.03	1750.12	0.992–1.090
Long x Reject	1.34	0.11	158.55	1.131–1.548

The analysis for the intermediate phase drew on the data for 'accept' gap-acceptance cases only. The analysis did not reveal any significant effects involving target vehicle and/or time gap.

The analysis for the completion phase also drew on just the data for the 'accept' gap-acceptance cases. The final analysis revealed a significant main effect of target vehicle ($\chi^2(1)$=7.03, p<0.01) and a significant target vehicle x time gap interaction ($\chi^2(4)$=190.62, p<0.01). The interaction is given in Table 5.2. There was a different pattern of effects depending on whether the PTW headlights were on or off. When the headlights were off, the time taken to travel through the completion phase of the turn did not differ significantly between the three time gaps. However, when the headlights were on, participants took significantly more time to travel through the medium and also the long time gap than the short time gap. These results suggest that PTW headlight status is important, but whether or not any advantage is to be gained during the completion phase of the turn is contingent on time gap.

Table 5.2   **Time gap x target vehicle interaction parameter estimates for the time taken to travel through the completion phase**

Parameter	B	SE	Wald chi-square	95% Wald confidence intervals
PTW lights on x Short	1.05	0.04	579.83	0.963–1.134
PTW lights on x Medium	1.28	0.06	422.72	1.154–1.397
PTW lights on x Long	1.27	0.04	946.92	1.193–1.355
PTW lights on x Short	1.31	0.11	139.25	1.092–1.527
PTW lights off x Medium	1.22	0.07	353.31	1.094–1.348
PTW lights off x Long	1.39	0.09	240.68	1.217–1.570

Safety Margin

The data for safety margin were taken from those trials where the target vehicle was a PTW either with headlights on or off. Calculated for 'accept' gap-acceptance cases only, the safety margin was defined as the distance, measured in seconds, between own-cab and the front of the target vehicle when own-cab was directly in the path of the target vehicle during the turning manoeuvre. In analysing the safety margin data, the question of interest was whether, at a given time gap, there was a difference in safety margin between target vehicle types – that is, PTW headlights on versus PTW headlights off. A separate analysis was carried out for each time gap. In every case, the final analysis showed that there was no significant effect of target vehicle on safety margin.

How Do the Findings Relate to the Research Questions?

To reiterate, the main question we asked in this experiment was:

> Is an oncoming vehicle that is a PTW with low-beam headlights on associated with a lower rate of gap acceptance, a more efficient turn strategy and larger safety margins when turning than an oncoming vehicle that is a PTW with low-beam headlights off?

The answer to this question is both 'yes' and 'no'. With respect to gap acceptance, target vehicle did have an impact, but this impact was dependent on time gap. Specifically, at the shortest, and most safety-critical, time gap only, a PTW with low-beam headlights on was associated with fewer gap acceptances than a PTW with low-beam headlights off. With regards to turn strategy, an effect of headlight status was restricted to just the completion phase. Statistically significant gains in turn time were found during the final phase of the turn and at the short time gap only. This outcome suggests that PTW headlight status does have an effect on turn time, but that this effect is more of a 'last-chance' one. In other words, at the shortest time gap only, when faced with the PTW with low-beam headlights on, drivers either rejected the gap or, in accepting the gap, adjusted for the smaller time gap by travelling more quickly to ensure safe completion of the turn. PTW headlight status did not significantly impact safety margin. With reference to our research question, a PTW with low-beam headlights on was not associated with a larger safety margin than a PTW with low-beam headlights off. Critically, in the case of each measure, there were no statistically significant occurrences of less safe performance when the headlights were on than when they were off.

What Are the Implications of the Findings for PTW Sensory Conspicuity and PTW Safety?

The results of the study are clear and provide evidence in support of low-beam headlights on PTWs as a way in which to enhance PTW sensory conspicuity during daytime conditions. With regards to gap acceptance, the outcome is consistent with the work of Olson et al. (1981) who also found that drivers accepted fewer gaps ahead of a motorcycle with low-beam headlights on than one with low-beam headlights off. However, the current work extends the work of Olson et al. in two important ways: (1) by demonstrating directly a role for time gap, and (2) by studying performance measures additional to gap acceptance – namely, turn time and safety margin.

The present research focused on just one type of sensory conspicuity treatment – that is, low-beam headlights. Thus, it is not known how the effects on turning decisions and performance observed here for motorcycle low-beam headlights translate to other sensory conspicuity treatments. Front lighting configurations and systems (for example, blinking lights) which aim to improve motorcycle discriminability through the provision of a unique visual signature (for example, Gershon and Shinar, 2013; Rößger et al., 2012) hold much promise, and so would be worth exploring for their effects on drivers' gap-acceptance decisions and performance.

From a safety perspective, the findings indicate improvements in PTW sensory conspicuity – in this case, through low-beam headlights – are associated with effects on performance that are reflective of a more cautious and risk-averse approach to driving by car drivers in the vicinity of oncoming PTWs. This is likely to translate to fewer crashes involving a passenger car and PTW at least for the scenario where the car is turning across the path of the PTW and the PTW has right-of-way. The currently underway case-control study of serious injury PTW crashes in Victoria, Australia (Allen et al., 2013) may, in due course, offer some insight in to this issue.

It is interesting to consider what these findings mean for understanding, at a more fundamental level, how sensory conspicuity treatments work. As noted at the outset, this experiment was based on the assumption that earlier detection of PTWs (that is, identification at longer distances from the PTW) due to heightened sensory conspicuity will lead to car drivers adopting safer driving practices when turning in the presence of PTWs. As we did not address this issue directly, we cannot say with absolute certainty that this was the case. Even so, the quickly expanding evidence-base highlighting a link between heightened PTW sensory conspicuity and earlier PTW detection certainly lends credence to this assumption. In a recent driving simulator study, Rogé et al. (2012), for example, showed that car drivers noticed from further away a PTW with high sensory conspicuity than a PTW with relatively low sensory conspicuity when the PTW was located in a position which was in front of the participant. Rogé et al. explained this effect from the point-of-view that, when sensory conspicuity is sufficiently high, the driver's attention is more readily captured and directed toward the more salient objects in the environment. This is indicative of a dominant bottom-up mechanism for detection at least. An aim of further research can be to explore systematically with a view to unpackage the precise mechanisms underlying the effects which were observed in the current experiment.

Experience as a Rider and Drivers' Turning Performance

In the experiment just described, our focus was on sensory conspicuity. However, other types of conspicuity have also been presented. Increasingly, more attention is being given to the concept of 'cognitive' conspicuity. While discussions on sensory conspicuity treatments tend to emphasize a role for bottom-up processes in PTW detection, interest in levels of 'cognitive' conspicuity highlights a primary role for top-down processes.

An object's cognitive conspicuity refers to the salience or prominence of an object as determined by its significance to the observer (for example, Engel, 1976). Wulf et al. (1989) have discussed that drivers who are also PTW riders have a heightened awareness of, and are more attentive towards, PTWs on the road, than are drivers who are not also PTW riders. Because of the low prevalence of PTWs on roads relative to passenger cars, drivers are not expecting to see a PTW, rather, drivers are expecting to see a passenger car. The implication is less efficient detection of PTWs. However, owing to their experiences, drivers who are also riders may be more expectant of PTWs and, and as a result, more primed for PTWs to be present. This has the potential to lead to non-compromised detection of PTWs.

To our knowledge, no research has been conducted to date to explore directly the role of cognitive conspicuity on drivers' decision-making and performance, including in situations when PTW riders are likely to be at their most vulnerable, such as when approaching an intersection in the path of a turning vehicle. Thus, in the experiment just described, we also sought to examine the influence of cognitive conspicuity. The amount of participants' direct experience as a PTW rider determined the level of cognitive conspicuity. There were two groups: 'drivers' – that is, experienced car drivers with no direct PTW riding experience; and 'driver-riders' – that is, experienced car drivers who were also experienced PTW riders.

Our question of interest was:

> For experienced car drivers, is being an experienced PTW rider associated with a lower rate of gap acceptance, a more efficient turn strategy and larger safety margins when turning across the path of a PTW than having no direct experience as a PTW rider?

The method and findings of this aspect of the experiment are described below. As above, a selection of these results has also been published elsewhere (Mitsopoulos-Rubens and Lenné, 2012).

What Did We Do?

Driving Simulator Trials and Procedure

The driving simulator trials and procedure were as described above.

Participants

The participants who took part in the experiment could be distinguished according to their direct experience as a PTW rider. Of the 43 experienced car drivers who participated, 20 (19 males and 1 female) were classified as 'driver-riders' and 23 (17 males and 6 females) as 'drivers'. The drivers had no direct riding experience – that is, they had never ridden or held a licence to ride a PTW. The driver-riders were all experienced riders – that is, they all held a current PTW licence and had held that licence for at least three years. A summary of selected demographic and driving experience information is given in Table 5.3 for the two groups separately.

Table 5.3   **Participant demographic and driving experience summary**

		Group	
Variable		**Driver**	**Driver-riders**
Age (years)			
	Mean	33.2	40.9
	Range	24–54	24–55
Years licensed as a driver			
	Mean	13.6	22.0
	Range	4–33	4–37
Years licensed as a rider			
	Mean	—	16.1
	Range	—	3–35
Hours driving per week			
	Mean	8.4	7.5
	Range	1–15	1–20
Hours riding per week			
	Mean	—	5.8
	Range	—	1–20

Analysis

Turning performance was defined, and each measure analysed, as described above.

What Did We Find?

Gap Acceptance

In the case of gap acceptance, the analysis sought to explore the effect of group on drivers' gap acceptance and, critically, whether group interacted with target vehicle to influence drivers' gap-acceptance rate. If cognitive conspicuity is indeed an important factor in drivers' decisions as to whether or not to accept a gap when turning across the path of oncoming traffic, driver-riders might be less likely than drivers to accept a gap ahead of a PTW, irrespective of headlight status, and with no difference between groups ahead of a car. The analysis did not reveal any significant effects involving the group variable.

Turn Time

As noted above, the data on turn time were taken from those trials where the target vehicle was a PTW, either with headlights on or off. For all gap-acceptance cases, the turn was divided into three phases: approach, intermediate and completion. All cases (that is, gap acceptances and rejections) were considered in the approach phase analysis, while only the gap acceptances were included in the analyses for the intermediate and completion phases.

A significant effect involving the group variable was found in the case of the intermediate phase only. Specifically, for the intermediate phase, a significant main effect of group was found ($\chi^2(1)$=9.37, p<0.01): when accepting the gap, the drivers took significantly more time to traverse the distance spanning the intermediate phase of the turn than did the driver-riders (Table 5.4).

Table 5.4   Group main effect parameter estimates for the time taken to travel through the intermediate phase

Parameter	B	SE	Wald chi-square	95% Wald confidence intervals
Drivers	2.07	0.06	1163.16	1.948–2.185
Driver-riders	1.80	0.06	867.74	1.680–1.920

Safety Margin

With respect to safety margin, of interest was whether, at a given time gap, there was a significant difference in safety margin between groups. In the case of the short time gap, the analysis did not find a significant difference between the groups. For both the medium and long time gaps, however, the final GEE model revealed a significant main effect of group (medium: $\chi^2(1)=10.56$, $p<0.01$; long: $\chi^2(1)=4.92$, $p<0.05$). As reflected in Figure 5.4, the mean safety margin of participants in the driver group was significantly shorter than that of participants in the driver-rider group. This was true at both the medium and long time gaps.

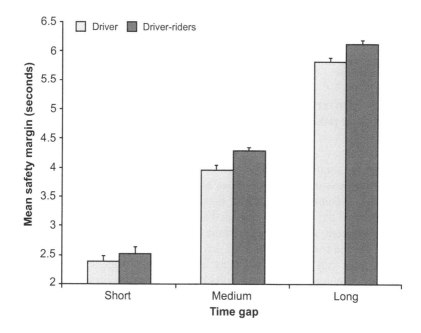

Figure 5.4   Mean safety margin (seconds) for each of the driver and driver-rider groups at each time gap

Note: Error bars represent +SEM.

Driver-riders' Awareness of Rider Crash Types

Driver-riders' awareness of the crash types which they are most at risk of experiencing as a rider may contribute to improvements in PTW cognitive conspicuity in specific scenarios as opposed to others. While we did not explore

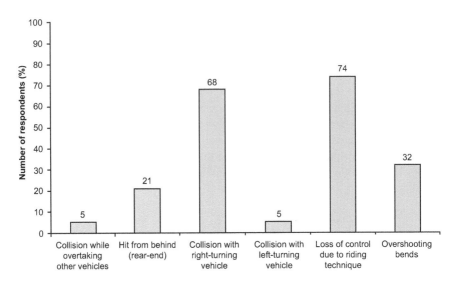

Figure 5.5   Driver-riders' perceptions of which crash types they are most at risk of experiencing as a ride

this issue in the simulator, we did collect some information from the driver-riders on their awareness of rider crash-types as part of the demographic and driving experience questionnaire. To provide some insight into their level of awareness, we administered the same question as that used by Clarke et al. (2004) in their UK-based in-depth study of PTW crashes. From a list of six crash types (four multi-vehicle and two single-vehicle), driver-riders were asked to indicate which two crash types they felt that they were *most* at risk of experiencing as a rider. Participants' responses are summarised in Figure 5.5. 'Collision with right-turning vehicle' was the multi-vehicle crash type selected most by the driver-riders (and second most overall). An awareness that PTW riders are at an elevated risk of crashing with a vehicle turning right may make drivers more cautious under these circumstances. Whether such an awareness translates into cognitive conspicuity-associated improvements in driving performance requires systematic investigation.

How Do the Findings Relate to the Research Question?

Our question here was:

> For experienced car drivers, is being an experienced PTW rider associated with a lower rate of gap acceptance, a more efficient turn strategy and larger safety margins when turning across the path of a PTW than having no direct experience as a PTW rider?

Our answer to this question is: 'it depends on the measure'. Riding experience was not found to affect gap acceptance, either overall or at any of the three time gaps under study. However, riding experience was found to have a significant impact on both turn time and safety margin. The effect of riding experience on turn time was restricted to the intermediate phase of the turn. For cases where the gap was accepted, the driver-riders traversed the distance immediately preceding the turn in less time than did the drivers. This effect was independent of time gap. Riding experience was also found to significantly affect safety margin. At both the medium and long time gaps, the safety margin adopted by the driver-riders was significantly longer than that of the drivers. This suggests that drivers with riding experience are negotiating the turn in such a way that translates to a larger gap between their vehicle and the oncoming PTW. This is most pronounced at the medium and long time gaps. The faster travel time during the intermediate phase among the driver-riders is in accordance with this finding.

What Are the Implications of the Findings for PTW Cognitive Conspicuity and PTW Safety?

The outcomes of our driver and driver-rider comparisons point towards riding experience as a mechanism through which to augment PTW cognitive conspicuity and, hence, PTW safety. While riding experience did not influence drivers' gap-acceptance performance in the current study, it did impact in a positive manner the way in which drivers negotiated the turn.

Albeit few in number, to our knowledge, investigations into PTW cognitive conspicuity to date have focused on processes surrounding PTW detection. Interestingly, in their driving simulator study, Rogé et al. (2012) found evidence of earlier PTW detection among their driver-riders than among their drivers, but this was for the situation where the PTW was positioned behind the participant. When the PTW was positioned in front of the participant, there was no difference in visibility distance between the two groups. While this latter finding might explain why, in the current study, we failed to observe an effect of riding experience on gap acceptance, it does not explain clearly why riding experience was found to impact other measures of turning performance. One assumption that we can make is that while top-down mechanisms supporting detection and perception may not flow through to support drivers' decisions about whether or not to accept a gap, top-down mechanisms may nonetheless come in to play later in the information processing sequence to support drivers' decisions and actions directly. This is seen in the form of more efficient turn time and larger safety margins. Of course, this is all speculative at this point. Further research would be needed to explore this assumption more definitively.

Of further note is that, in the current research, we did not distinguish between different degrees of riding experience. That is, we did not compare the performance of our drivers and driver-riders with drivers who, compared with our driver-riders, have only a small amount of riding experience. Therefore, we do not know how much riding experience is sufficient in order to realize performance benefits associated with improvements in cognitive conspicuity. This is an area worthy of further enquiry, particularly given the potential safety gains that could come from including a discussion in driver-training programs and motorcycle safety awareness-raising campaigns of the specific risks faced by riders.

Conclusion

Overall, the work described here provides support for the use of low-beam headlights and riding experience as tools through which to augment the sensory and cognitive conspicuity of PTWs, respectively. Further research in a more realistic setting would be worthwhile in helping to validate the present findings, and in helping to explore more definitively the precise mechanisms underlying the effects observed here. Ultimately, we would hope to see improvements in PTW conspicuity translate to fewer crashes involving PTWs under those circumstances where diminished conspicuity would have played a critical role.

Acknowledgements

Financial support for the current research was provided through a National Health and Medical Research Council-European Union Health Collaborative Research grant. The authors would like to thank the following colleagues for their involvement: Ms Miranda Cornelissen, A/Prof Stuart Newstead, Mr Nicolas Reid, Prof Geoff Underwood, Mr Ashley Verdoorn, and Ms Amy Williamson.

References

ACEM. (2009). MAIDS: in-depth investigations of accidents involving powered-two-wheelers (Final report 2.0). Accessed 16 March 2010 from www.maids-study.eu/pdf/MAIDS2.pdf

Allen, T., Day, L., Lenné, M., Symmons, M., Newstead, S., Hillard, P. and McClure, R. (2013). Finding evidence-based strategies to improve motorcycle safety: A case-control study on serious injury crashes in Victoria. *Proceedings of the 2013 Australasian Road Safety Research, Policing and Education Conference.* Brisbane, Australia.

Caird, J.K. and Hancock, P.A. (2002). Left-turn and gap acceptance crashes. In Dewar, R.E. and Olson, P.L. (eds), *Human Factors in Traffic Safety*, 613–52. Tuscon, AZ: Lawyers & Judges.

Clarke, D.D., Ward, P., Bartle, C. and Truman, W. (2004). *In-depth Study of Motorcycle Accidents* (Report No. 54). Department for Transport, London, UK.

Clarke, D.D., Ward, P., Bartle, C. and Truman, W. (2007). The role of motorcyclist and other driver behaviour in two types of serious accident in the UK. *Accident Analysis and Prevention*, 39, 974–81.

Cole, B.L. and Jenkins, S.E. (1984). The effect of variability of background elements on the conspicuity of objects. *Vision Research*, 24, 261–70.

Engel, F.L. (1976). *Visual Conspicuity as an External Determinant of Eye Movements and Selective Attention.* Unpublished doctoral thesis. University of Technology, Eindhoven, The Netherlands.

Gershon, P. and Shinar, D. (2013). Increasing motorcycles attention and search conspicuity by using Alternating-Blinking Lights System (ABLS). *Accident Analysis and Prevention*, 50, 801–10.

Hole, G.J., Tyrell, L. and Langham, M. (1996). Some factors affecting motorcyclists' conspicuity. *Ergonomics*, 39, 946–65.

Itti, L. and Koch, C. (2000). A saliency-based search mechanism for overt and covert shifts of visual attention. *Vision Research*, 40, 1489–506.

Liang, K.Y. and Zeger, S.L. (1986). Longitudinal data analysis using generalised linear models. *Biometrika*, 73, 13–22.

Mitsopoulos-Rubens, P. (2010). *Calibration Ability and the Young Novice Driver.* Unpublished doctoral thesis. Monash University, Clayton, Victoria, Australia.

Mitsopoulos-Rubens, E. and Lenné, M. (2012). Issues in motorcycle sensory and cognitive conspicuity: The impact of motorcycle low-beam headlights and riding experience on drivers' decisions to turn across the path of a motorcycle. *Accident Analysis & Prevention*, 49, 86–95.

Olson, P.L., Halstead-Nussloch, R. and Sivak, M. (1981). The effect of improvements in motorcycle/motorcyclist conspicuity on driver behaviour. *Human Factors*, 23, 237–48.

Rogé, J., Douissembekov, E. and Vienne, F. (2012). Low conspicuity of motorcycles for car drivers: Dominant role of bottom-up control of visual attention or deficit of top-down control. *Human Factors*, 54, 14–25.

Rogé, J., Ferretti, J. and Devreux, G. (2010). Sensory conspicuity of powered two-wheelers during filtering manoeuvres, according the age of the car driver. *Le Travail Humain*, 73, 7–30.

Rößger, L., Hagen, K., Krzywinski, J. and Schlag, B. (2012). Recognisability of different configurations of front lights on motorcycles. *Accident Analysis and Prevention*, 44, 82–7.

Transport Accident Commission. (2013). Motorcycle crash data. Accessed 16 August 2013 from www.tac.vic.gov.au/road-safety/statistics/summaries/motor cycle-crash-data

Wertheim, A.H. (2010). Visual conspicuity: A new simple standard, its reliability, validity and applicability. *Ergonomics*, 53, 421–42.

Wulf, G., Hancock, P.A. and Rahimi, M. (1989). Motorcycle conspicuity: An evaluation and synthesis of influential factors. *Journal of Safety Research*, 20, 153–76.

Chapter 6

Design Studies on Improved Frontal Light Configurations for Powered Two-Wheelers and Testing in Laboratory Experiments

Lars Rößger, Jens Krzywinski, Frank Mühlbauer and Bernhard Schlag

Thinking about Frontal Light Design – An Approach Promising a Safer Interaction between Riders and Drivers?

According to findings from accident studies (for example Williams and Hoffmann, 1979; ACEM, 2004; McCarthy et al., 2007) inadequate perception of powered two-wheelers (PTW) by other road users is one key factor in the causation of crashes between PTW riders and other motorists. In order to enhance the (visual) conspicuity of PTWs for other road users, numerous conspicuity measures, for example fluorescent clothing, frontal lighting, bike and helmet colour, side reflectors at the bike and/or the tires etc.and so on, have been considered within evaluation studies with respect to their safety benefits (an early comprehensive overview is given by Wulf et al., 1989) and are partly recommended by safety authorities and rider associations. In the present chapter, we focus on the frontal light configuration of PTWs as a potential conspicuity aid which might tackle some disadvantageous attributes of a PTW in the context of its perception by others. From our understanding, frontal light configuration refers to the arrangement, composition and the design of light sources within the frontal view of the PTW. That implies that a light configuration enfolds not only one single light source, for example a solo headlight, rather than a combination of several light sources, for example a solo headlight and additional position lights shaping up a triangle, as a whole.

Why do we assume that the frontal light configuration might contribute to a better perception, and, thus, to a safer interaction between riders and other motorists? A first point seems to be obvious: the enlargement of the luminance area by additional light sources potentially makes the PTW more visually salient within the road scene. Several studies indicated that the detectability of an object depends, among others, on its size and its luminance in relation to a given background (for example Wulf et al., 1989; Rogé et al., 2012). This is in line with bottom-up theories of visual attention: according to these the visual attention is directed by low-level features of the objects (such as contrast, orientation etc. and so on) to regions of high visual saliency within the visual scene (for example

Itti and Koch, 2000). Underwood et al. (2011) demonstrated that the decisions of subjects during the inspection of roadway scenes were clearly affected by the visual saliency of the critical vehicle, corresponding with riskier decisions in scenarios where low visually salient motorcycles were present in the road scene. These results were accompanied by findings from eye-tracking analyses indicating that high, visually salient region earlier attract the attention and that the identity of motorcycles, particularly under the low salient condition, were more difficult to determine by the participants.

A second point refers to the idea that a frontal light configuration might convey a unique signal pattern facilitating a fast and easy recognition and identification of PTWs in the road scenery. In the context of the progressive introduction of daytime-running lights (DRL) on cars (ECE R48/87 legislation), several researchers expressed concerns that the conspicuity of PTWs might be additionally reduced due to this developement and considered a distinctive illumination for PTWs as a possible measure to counteract this reduction and to make PTWs more easily identifiable (Koornstra et al., 1997; Cavallo and Pinto, 2012; Thomson, 1980). In the same way, Rumar (2003: 25) argued that one potential benefit of a special DRL configuration on motorcycles could solve problems of their late identification by others. That is, a unique and clearly distinguishable frontal light configuration, aptly termed by Cavallo and Pinto as some kind of a specific visual signature, could provide a distinctive feature for shortening recognition processes during the traffic-scan operations by road users. Hole and Tyrrell (1995) conducted a study which provides some empirical support for this assumption: the authors exposed subjects to a series of slides depicting road way scenes with various vehicle types and obtained subjects' detection time for detecting motorcycles. As between-factors they varied: (a) the motorcycle's headlight status (ON vs OFF) and (b) the consistency of the headlight status in the final slide related to the preceding slides. Results showed that the detection of the motorcycle in the final slide was significantly slower if a motorcycle with unlit headlights was depicted after the repeated presentation of motorcycles with headlights on. The authors reasoned that drivers, based on their experiences, develop what they called 'perceptual set': simple perceptual features will be associated with the presence of a motorcycle, and, in consequence, will be used as a cue to code the presence of a motorcycle (see also Hole, 2007). The advantages are obvious: cue-driven recognition would disburden and shorten recognition/decision processes, especially in situations with time pressure. However, problems, for example delays or failures in the detection, will arise if these cues become inconsistent, as the results reported above suggest, or if they become indistinctive. The latter might be potentially induced by the introduction of car DRL (Cavallo and Pinto, 2012).

A third point refers to potential effects of frontal light configuration on drivers' speed estimations regarding to PTWs. The misjudgements of PTWs' speed by other motorists is seen as one critical factor contributing to the PTW riders' crash risk. The size-arrival effect, as one prominent explanation describes, is that small objects result in a smaller retinal size and are thus estimated to arrive later than retinally large objects (DeLucia and Warren, 1994; Horswill et al., 2005). This might have severe implications for roadway decisions, for example whether I should pull out in front of a vehicle or should I wait until the vehicle has passed by? Their relative small front and irregular outline exacerbate the estimation of PTWs' speed, lead to an overestimation of their arrival time and are thus, in consequence, more likely subject to safety-critical decision outcomes. Results from Gould et al. (2012a) indicated that in order to perceive a motorcycle and a reference vehicle (car travelling at 30 mph) as equally fast, the actual approach speed of the motorcycle was significantly faster related to the reference vehicle. Particularly under night-time conditions, their study revealed drastical results: the threshold for an equal speed perception relative to an oncoming car at 30 mph was on average 146 mph (!) for the motorcycle. This points to the issue that the frontal lighting as a significant perceptual element in order to judge vehicles' speed and distance is disadvantageous for PTWs because PTWs' front often contains just one headlight or dual headlights which are placed very close together. Studies focusing on the impact of the width of headlight separation on distance estimations showed that decreased inter-headlight width goes along with subjects' tendency to overestimate the distance of a vehicle (Castro et al., 2005). Further studies indicated that an improved headlight design on motorcycles might positively affects drivers' choice of safety margins during turning manoeuvres (Tsutsumi and Maruyama, 2008) and the accuracy of speed judgements of oncoming motorcycles (Gould et al., 2012b).

The Design Process: Search for Signal Patterns and their Implementation

Concept Development by Design

The initial stage of the design work focused on the development of signal patterns which are potentially applicable to all PTWs as market and after-market equipment. With respect to their differences in style, riding performance and handling, it appeared to be necessary to develop different light configurations for two of the main categories of PTWs: motorcycles and scooters. Riding performance as well as riding style and behaviour within those two categories often differ, for example in maximum and average speed, acceleration and riding areas. If the aim is to develop a specific visual signature for a vehicle category which is supposed to be associated with homogeneous observations and thus with consistent expectations by other road users, scooters and motorcycles must be considered separately. As stated above, the frontal light arrangement covers the ensemble of common headlights and optional position lights as well as the coordination, integration and interaction of those components. In a geometric sense, there are four types of light arrangements: independent lights, grouped lights, combined lights and integrated lights.

The initial design sketches (see Figure 6.1) dealt with the front analysis of a motorcycle in order to find potential locations for additional light sources.

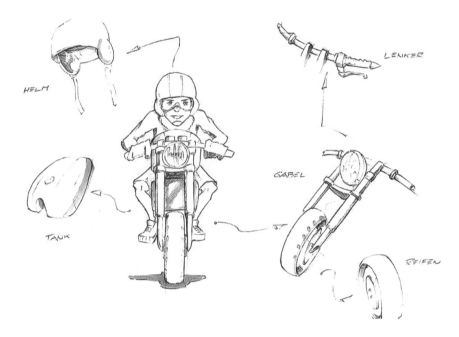

Figure 6.1   Design sketches for additional light sources

Employing the helmet and tires as light carrying parts appeared to be challenging even though great distances between the single light sources could be implemented. As a result of further investigations, the fork and handlebar seemed to be more feasible. Therefore, the pattern search started with those two parts. From these first sketches, three basic patterns with linear lights were derived (see Figure 6.2a). The first pattern consists of a horizontal light arrangement which illuminates the entire width of the handlebar.

Although this leads to a precise definition of the overall width, the height of the motorcycle cannot be determined. With the second pattern, the opposite is achieved. At this stage of the design, lights on both sides of the fork and an additional helmet light were used to create a vertical arrangement. Advantages of both the horizontal and vertical signal patterns were combined in the third approach. As a result, a T-shaped pattern was realized by using light sources on the handlebar as well as lights mounted on the fork. Thus, the motorcycle's total height and width are well represented and can be recognized as a prototypical motorcycle front.

With regard to the signal pattern of a scooter, the second attempt aims to find similar solutions by using punctual light sources. Thus, three light patterns were developed. The various configurations can be distinguished by the types of shape: X, Y and V (see Figure 6.2b). In the following design process, the Y-light arrangement is not further elaborated as it is very similar to the T-arrangement.

a) Signal patterns with linear lights sources   b) Signal patterns with punctual lights sources

Figure 6.2   Signal patterns of the concept phase

Therefore, only the X and V arrangements are considered further. On this basis, the abstract sketches were replaced by realistic images and definite street scenarios. To find the final signal pattern, two criteria were introduced in the design process: the appearance in front and side views on the one hand and the authentic description of the overall size on the other hand. The evaluation of all signal patterns and their combinations led to the decision to develop further both the V-arrangement for scooters and the T-arrangement for motorcycles. From a technical perspective and from the viewpoint of design, the following advantages and disadvantages are to be considered:

- The benefit of the V-arrangement (see Figure 6.3a) is the creation of a characteristic signal pattern in front and side views. Furthermore, this shape is appropriate for placement on single and double arm forks. Based on the lowest light position of the V-arrangement, its comparatively high position above street level is considered a weakness.
- The major advantage of the T-arrangement (see Figure 6.3b) is the appearance of the front view. Here, the abstract motorcycle shape with

Figure 6.3   Results from the concept phase – light arrangements for scooter and motorcycle

its overall dimensions becomes clearly identifiable. Furthermore, the linear light stripes appear larger compared to punctual light sources. A disadvantage, however, is its attachment solely to the double arm forks.

Along with the steps to define signal patterns for the frontal light configuration, different light sources, colours of light and light modulations were analysed to outline possible ranges for subsequent steps within the design process. However, not all possibilities seemed promising enough to pursue, as a permanent white light, preferably emitted by a light-emitting diode (LED), is still most beneficial.

Designs for Two Prototypical Powered Two-Wheels

To integrate these concepts in existing and forthcoming PTWs, two prototypical PTWs – a motorcycle and a scooter – were defined as reference. The KTM Super Duke was chosen as an explicitly aggressive naked bike with a facet surface and an additive styling. Being a benchmark of the scooter market, the Vespa GTS 125 was elected because in comparison to the KTM it represents a contrary design language with its smooth and well-integrated styling. As these two prototypical examples cover a wide range of PTW types, the T-arrangement was used with the KTM Super Duke, whereas the V-arrangement was used with the Vespa GTS.

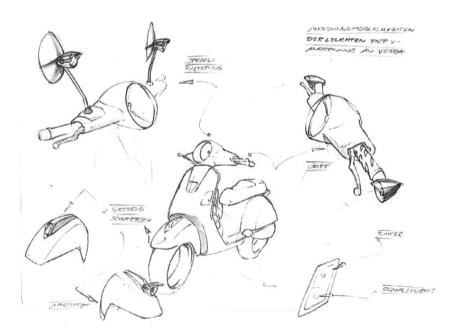

Figure 6.4   Options for additional light elements on a scooter

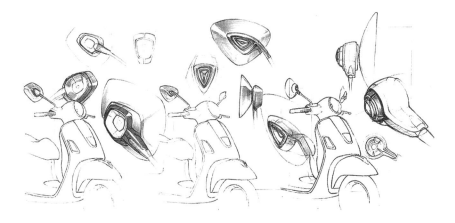

Figure 6.5   Sketches for different design concepts of the Vespa

Vespa GTS 125: not many possibilities existed for positioning the lights at the front end since the V-arrangement required five quite specific luminous points (see Figure 6.4). Hence, one light was mounted on the fender, two were integrated in the housing of the direction indicators within the front shield and the last two were mounted on the backside of the mirrors. The following sketches and renderings dealt with the implementation of the mirror lights, as the standard enclosures of the Vespa did not provide enough space for an additional light assembly. Therefore, the main task here was to develop a new set of mirrors which simultaneously represents the design language of the central light mounted on the fender. A new design did not have to be considered for the lights inside the housing of the direction indicators, as they fit into the existing construction space. In this context and at this stage, three different concepts were developed (see Figure 6.5).

KTM Super Duke: based on the design development of the Vespa, the KTM was investigated in an analogous manner. At first, all possible positions for the lights in a T-arrangement were examined (see Figure 6.6). The lower lights were either positioned in a linear alignment on the fork or integrated into the front fender. Furthermore, the integration of the upper lights inside the rear mirror housings or into elements of the handle bar seemed to be suitable. An outstanding detail of the KTM Super Duke is the design of the fender, which covers large parts of the fork. Thus, it is ideal for the integration of the lights. The various renderings (see Figure 6.7) focused explicitly on the integration of the linear lights, taking into account the formal and technical requirements. The decision to adopt this approach predetermined the use of horizontal lights integrated into the rearview mirrors. After finishing the initial drawing and rendering process, the first designs were transferred to five different scaled foam models, full-scale paper models and 3D renderings.

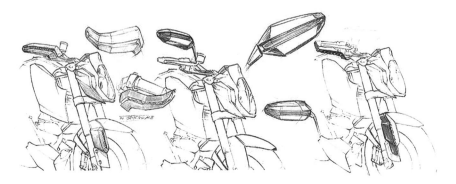

Figure 6.6   Options for additional light elements on a motorcycle

Figure 6.7   Sketches for additional light elements in a T-arrangement

Later, the design process focused on the KTM Super Duke, because the integration of linear lights was more complicated than the integration of punctual light sources. The final design solution for the Vespa was developed by means of the experience gained during the KTM design process. The design development ended with the generation of photorealistic renderings of a KTM Super Duke with a T-arrangement and a Vespa GTS 125 with a V-arrangement as seen in Figure 6.8.

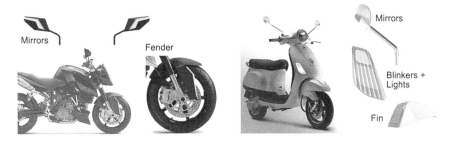

Figure 6.8   KTM Super Duke with T-light configuration and Vespa GTS 125 with V-light configuration

Figure 6.9   Design renderings of night-time conditions

Both designs show that the transformation of the former conceptual signal patterns of the T- and V-arrangements work well on two very different PTWs (see Figure 6.9a and 6.9b). This means that on the one hand geometric requirements can be met with linear or punctual lights positioned on real motorcycles and that on the other hand an integration of those light arrangements is easily possible. These characteristics show that safety elements can be implemented without causing disruption, which is a basic requirement for achieving acceptance among PTW riders. By integrating LED assemblies into finish models, technical assumptions about the light arrangements can be reviewed in the next steps.

Finish Modelling and Implementation

The first step of modelling contained the analysis of the present CAD models with basic surfaces and their suitability for the realization of full-scale prototypes. The second step included the scanning of the reference surfaces of the PTWs which enclosed the fender of the KTM and the front shield of the Vespa to gain knowledge of the geometric constraints for their later integration. Those data constituted a basis for the modelling of all necessary elements (see Figure 6.10).

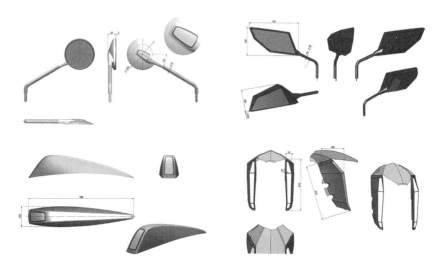

Figure 6.10   Solid CAD models for mirrors, fender and centre light

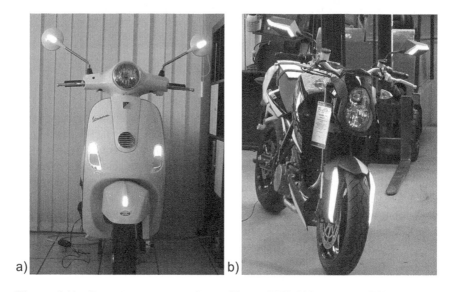

a)   b)

**Figure 6.11   Prototypes mounted on a Vespa GTS 125 and on a KTM
              Super Duke**

At this stage, all inner fixtures such as LEDs, cables and the control board as well as all connection points for later mounting on the PTWs were considered.

Based on the CAD models, the housings of the DRL components were built as hybrid prototypes combining semi-finished metal parts, different plastic parts and rapid prototyped elements. The chassis were equipped with the LED modules and closed by a translucent face plate. A control and power supply unit completed the prototypes. To test the prototypes, we mounted them on a KTM Super Duke and a Vespa GTS 125 (see Figure 6.11a and 6.11b). After demounting the final prototypes, they were scanned for additional use in virtual environments in further steps of the project.

Experimental Studies on Effects of Motorcycles' Frontal Light Configuration

In a series of laboratory experiments we tested effects of motorcycles' frontal light configuration (T-light configuration) on attention conspicuity, their identification by others and roadway decisions.

Experiment 1 – Method

A first experiment (for further details see Rößger et al., 2012) was conducted using a set of still pictures (photographs) depicted on a computer screen. A set contained 44 photographs and represented real traffic scenarios at intersections. During the experiment, participants had been instructed to identify as fast as possible all relevant vehicles they had to pay attention to when they were planning to pull out into the main road or, respectively, to cross the intersection. That is, subjects were not sensitized in advance to search especially for motorcycles (for methodological considerations on search conspicuity vs. attention conspicuity see Gershon and Shinar in this book). In order to furthermore avoid unrealistic expectations, a motorcycle, either alone or together with other road users, was depicted in 10 photographs (test slides) whereas other motorists (cars, public transport) without a motorcyclists were represented at the remaining (reference slides).

Two between-factors have been varied: a) the light configuration of the motorcycles' (control condition: solo headlight, experimental condition 1: T-light-configuration, experimental condition 2: T-light-configuration plus additional lights mounted at the rider's helmet), and b) – (according to Hole and Tyrrell, 1995) – the consistency of MC's light configuration in the final slide (inconsistent vs. consistent). The subjects (N=56, mean age=22.4, SD=5.1) were randomly assigned to the control group and to the experimental groups. We measured the time subjects required for identifying the relevant vehicles and gaze parameters when inspecting the intersection scenarios.

Experiment 1 – Results

Referring to the reference slides, the results did not reveal significant differences between the three conditions in the time subjects needed for identification (TI). This finding suggest that the randomized assignment to the conditions was successful and potential effects ascribable to the experimental manipulation of the independent variables. For the test slides, we could identify an effect of the light configuration on TI: particularly when the motorcycle was competing for visual attention due to the presence of other motorists, subjects identified relevant vehicles faster under the experimental condition than under the control condition (F2,49=3.746, p ≤ 0.05, see Figure 6.12, top). For the final slide – contrary to our assumptions and to the results from Hole and Tyrrell (1995) reported above – we did not find differences in TI when the scenario contained a motorcycle with solo headlights which was consistent with previously shown motorcycles compared to the scenario with a motorcycle with solo headlights which was inconsistent with the preceding motorcycles. That is, subjects did not need more time for the scene depicting a motorcycle with solo headlight when the previously presented motorcycles had an enhanced light configuration compared to the repeated presentation of motorcycles with solo headlights in advance. However, comparison between the presentation of solo headlight vs. enhanced light configuration in the

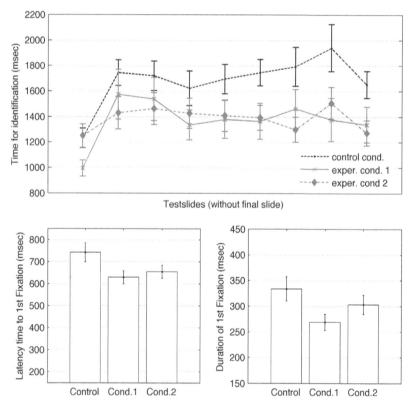

Figure 6.12   Means and standard errors: time for identification (top), latency times and duration for 1st fixation (bottom)

final slide within both experimental conditions hinted also at tendencial shorter TI for the scene containing a motorcycle with enhanced light configuration.

Referring to gaze parameters we analysed the time to the first fixation at the target (motorcycle) and the duration of the first fixation at the target (see Figure 6.12, bottom). Parameters were subjected to an ANOVA with the motorcycle's light configuration as between-factor. Results revealed marginally significant main effects of the light configuration on the time to the first fixation and on the duration for the first fixation. *Post-hoc* tests for multiple comparisons indicated that subjects fixated more quickly on a motorcycle equipped with T light configuration (condition 1) compared to the solo headlight (control condition) and that subjects need shorter fixation duration under condition 1 compared to the control condition (Fisher LSD; p=0.04, and respectively, p=0.03).

The observation of shorter times to first fixate a motorcycle with an T- light-configuration suggests that the visual attention of the subjects was earlier attracted by that kind of light configuration compared to motorcycles with solo headlight configurations. That is in line with assumptions based on models of visual attention accentuating visual low-level features and the role of bottom-up processes for directing the attention. The additional light sources enhanced the luminescent area within the motorcycle's front, and, thus, strengthened its contrast to the background. Notwithstanding, they used other indicators (for example length of saccadic movement, number of fixation prior to the target fixation). Underwood et al. (2011) reported similar results as they found that the visual saliency of vehicles affects the gaze behaviour, corresponding with faster attraction of attention by high visually salient vehicles. The consideration of first fixations' durations furthermore suggest that the visual processing of the motorcycle was easier to succeed when a motorcycle with T-light configuration was presented. The results from analyses of the gaze behaviour were associated with shorter time spans the subjects needed to identify the relevant vehicles in the experimental conditions. It remains, however, somewhat unclear whether this effect was triggered due to the increased visual prominence of this light configuration or because the light configuration as a concise visual signature served a cue for shortening recognition processes. First assumption is supported by results by Underwood et al. showing that low saliency vehicles received longer inspections by subjects than high saliency vehicles. Referring to the latter hypothesis, the comparison between consistent and inconsistent presentation in the final slide did not show differences in the time required for identification. That implies that the specific visual signature failed to become a shorthand cue in order to detect the presence of a motorcycle in our experiment. One reason for the missing effect might be that the subjects could not develop a 'perceptual set' because of the briefness of the preceding experiences phase, and, respectively, the relative low frequency of motorcycles depicted in our set compared to other studies. Hole (2007) also noted that the size of the effect by a potential signal code for motorcycles is affected by the frequency of presented motorcycle (associated with the signal) in advance.

It has to be mentioned that the experiment has some critical points from a methodological point of view. First, the presentation of light sources seems to be critical because of the restricted luminance of computer screens. One might assume that the effects under real conditions with real light sources might be even stronger, nevertheless a validation of our results is needed before general conclusions are drawn. Second, still pictures represent the situations in real traffic just partly because of their static characteristics: further attention-capturing information which seems highly relevant in a dynamic traffic scenario such as strong motion transients (Franconeri et al., 2005) created by vehicles are neglected. There is, however, some empirical evidence suggesting that results originated by experiments using static scenes can be transferred at least to some extent also to dynamic scenes (see Underwood et al., Chapter 4 in this book). In a second experiment we considered the effects of a motorcycle's light configuration within a dynamic scenario, albeit the variable of interest here refers rather to the appraisal of a motorcycle's speed than to its attention-capturing capability.

Experiment 2 – Method

Experiment 2 considered the effect of the T-light configuration on judgements and decisions drivers make at intersections. As stated above, road users' difficulties in estimating speed and distance of approaching PTWs is seen as one critical issue related to PTWs' accident risk. We furthermore were interested whether and how the simultaneous presence of other vehicles affects drivers' decision-making towards motorcycles in left-turn scenarios. Perceived time-to-collision (TTC) as an essential underlying variable for such a decision process has been investigated in numerous studies for single objects (for comprehensive overview see Hecht and Savelsbergh, 2004). Baurés et al. (2010) noticed that comparatively little attention has been given as yet to TTC judgements for multiple objects. The authors examined the performance of two parallel TTC judgements (leading and trailing object) and found that the TTC for the later-arriving object was systematically overestimated while the TTC judgement for the first-arriving was unaffected. The authors argued that the effect was caused by proactive interferences induced by estimating the arrival of the first stimuli in a dual task paradigm. Another experiment focused on TTC-related judgements regarding a vehicle on a collision course with the subject whereby a second, task-irrelevant distractor-vehicle was presented on a parallel trajectory (Oberfeld and Hecht, 2008). Results showed that the TTC for the task-relevant object was systematically underestimated when the distractor arrived later than the task-relevant object.

A two divided simulator study examined gap-acceptance behaviour in a left-turn scenario using a critical time gap task (CTG). Gap acceptance refers to a driver's decision about whether a gap ahead of an approaching vehicle is either acceptable or unacceptable to turn in front of the vehicle. The CTG describes the time gap between the observer and the oncoming object at the moment the driver chooses to reject the gap in a gap-acceptance paradigm. That is, the greater the time gap the greater the chosen safety margin.

The simulator study was two divided as we implemented the simulation in two different simulation environments and each participant took part in two sessions. That allows not only to examine the consistency of potential effects across different simulations modes *per se* but also to test whether the effects depend on differences in luminous intensity during the simulation. Within session 1, the experimental task was implemented in a five-sided 3D-Computer Animated Virtual Environment (CAVE). The participants are placed in a room where all walls are used as presentation planes except the wall in the participant's back. The virtual environment was continuously adapted according to participant's head position by an infra-red tracking system. This simulation enables participants to completely immerge in a virtual 3D road way scene. In session 2, the simulation was presented at a 3D-Power wall. The 3D-Power wall projected the scene on a single plane in front of the participants, whereby this presentation was characterized by a higher luminous intensity compared to session 1.

During both sessions, the virtual scenery depicted a T-junction with multiple traffic lanes (two lanes for each direction for the main road and one lane for each direction for the secondary road). The participant sat in a virtual cockpit of a car (EGO vehicle) with the intention to turn left from the main road into the secondary road and had to pay attention to approaching vehicles from the opposite direction. From the opposite direction either a single motorcycle approached (always on the outer lane) or a motorcycle and a red car (always on the inner lane) approached the intersection. The oncoming vehicle(s) started in a distance of 160 metres to the participant. The motorcycle approached the intersection with a velocity of v=16.7 m/s (60 km/h) and the car with v=13.9 m/s (50 km/h). That is, when a trial contained two oncoming vehicles both vehicles started at the same time and the motorcycle arrived at the intersection earlier than the car.

As within-factors we varied the light configuration of the motorcycle (solo headlight vs. T-light configuration) and the presence of the distractor vehicle (with car vs. without car at parallel trajectory), resulting in 2 x 2 design with repeated measurements. In each session each condition was repeated four times and the order was randomized within four blocks. The frontal light configuration was modelled within the software 'EON Reality' for the 3D scene visualization. Because of the unavailability of self-luminous surfaces, the flood light consisted of a translucent, parabolic light cone geometry with a punctual light source in the focal point. For the T-light configuration this light assembly was repeated, transformed and placed at the fork legs and rearview mirrors. According to a balanced assignment of the sample to a starting session, the participants started either with session 1 (3D CAVE) or with session 2 (3D Power wall).

Experiment 2 – Results

A total of 39 participants (20.5 per cent female and 79.5 per cent male) completed the study. Participants' age was M=30.7 years (SD=5.63) and ranged from 23 to 46 years. All participants held a valid car driving licence and reported normal or corrected-to-normal vision.

For session 1, the mean for CTG was 4.978 msec (SD=1.251), and for session 2 respectively, the mean CTG mean was M=5.253 msec (SD=1.1615). CTG values were analysed with two-factor ANOVA for repeated measurements, separately for the two sessions.

For session 1 (Figure 6.13 left), the main effect of motorcycle's front light configuration yielded an F ratio of F(1,37)=97.98, p < 0.01, corresponding with a greater GTC (M=5.187 msec, SD=1.140) and indicating that subjects earlier rejected the turn manoeuvre when a motorcycle with T-light configuration approached the intersection than when the oncoming motorcycle had a solo headlight (M=4.770 msec, SD=1.165). The main effect of car presence yielded an F ratio of F(1,37)=6.674, p=.014, corresponding with a greater GTC (M=5.025, SD=1.140) and with earlier decisions to give up the turn manoeuvre when no car was presented than under the condition a slower car was presented parallel to the motorcycle (M=4.932, SD=1.1589). The interaction effect was non-significant, F(1,37)=0.472, p=.496.

For session 2 (Figure 6.13, right), results revealed again a significant main effect of motorcycle's front lighting on the time to give up the turn manoeuvre (F(1,38)=9.398, p ≤0.01), corresponding with a greater GTC (M=5.320, SD=1.117) and indicating that participants earlier decided to give up a turn manoeuvre when the motorcycle was equipped with a T-light configuration compared to a solo headlight at the motorcycle (M=5.186, SD=1.2178). The main effect of car

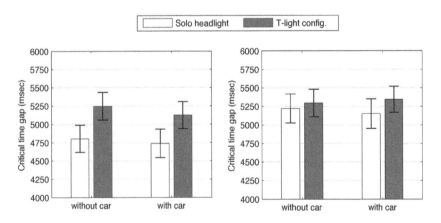

Figure 6.13   Means and standard errors for critical time gaps (GTC) in session 1 (left) and session 2 (right)

presence was non-significant ($F(1,38)=0.076$, $p=.784$). However, the interaction effect was significant, $F(1,38)=4.672$, $p \leq .05$. Focusing on scenery where the approaching motorcycle had the solo headlight configuration, participants decided earlier to reject the gap when the car was not present compared to the condition with car. The reverse picture emerged when the motorcycle was equipped with the T-light configuration: here participants decided earlier to give up a turn manoeuvre when a car was present compared to the condition without car.

The results suggest that an enhanced frontal light configuration like the T-light configuration potentially provide some safety benefits referring to drivers' appraisal of approaching motorcyclist. The finding of increased safety margins, that is, increased time gaps between observer and motorcycle, was consistent across two experimental sessions. Furthermore, results are consistent with findings on the accuracy of speed judgements towards motorcycles (Gould et al., 2012b) and results originated from real-world studies (Tsutsumi and Maruyama, 2008). The observation that the effect was greater under lower luminance conditions (session 1) suggests that this safety benefit might be stronger when the surrounding brightness is reduced, such as during twilight, dawn or night-time conditions. That seems plausible as the light sources at the motorcycle's front apparently become even more significant perceptual elements during these times for judging a motorcycle's speed and distance.

That the presence of a slower car at a parallel trajectory to the motorcycle negatively affects the safety margin is inconsistent with previous studies (Oberfeld and Hecht, 2008). Based on their findings of systematic underestimation of the TTC of a target vehicle when an distractor vehicle was arriving later, one might assume that the time gap between observer and target vehicle in the GTC paradigm should be increased. This inconsistency might be caused due to the difference in distractor's task-relevance: the distractor presented by above-mentioned authors was task-irrelevant as the vehicle was not on a collsision course with the observer, whereas in our experiment the distractor vehicle became task-relevant as the observer had to consider which of both objects would arrive at first. In this respect, our findings provide some support for the hypothesis by Oberfeld and Hecht that the task-related significance of a distractor will determine the way TTC information is processed to a final TTC judgement.

Conclusion

Our studies provided some promising results referring to potential safety benefits of frontal light configuration design on PTWs. So, visual attention of observer was attracted earlier by a PTW with enhanced light configuration. Furthermore, the visual processing and identification of the relevant vehicle in intersection scenarios was more quickly executed. De Graen et al. (2011) critically noted that the effects might caused by some kind of novelty/surprise effect due to the uncommon appearance of the PTW. We indeed found prolonged fixation durations for the first exposure (related to the control condition), suggesting the existence of a novelty effect; subsequent presentations however did not receive a level equal to the control condition, but shorter fixation durations. That is somewhat contradictive to the assumption that the effects are solely based on observers' surprise with an ensuing decrease to the baseline. But we agree that further research and validation is needed before generalized conclusions can be drawn. That is also true in the context that we could not prove doubtless that faster identification of PTWs rely on the distinctive visual signal/cue conveyed by PTWs' frontal lighting. As the argumentation and the empirical findings by Hole and Tyrrell (1995) seem however very cogent, the absence of supportive effects in our study might be caused by methodological shortcomings referring to that issue. Apart from the findings on attention and identification, we found also some empirical support that an enhanced light configuration could facilitate the PTWs' speed appraisal by other road users.

With respect to actual safety benefits, we have to mention possible barriers: first, mechanism of risk compensation by riders might lead to (over-)compensatory, riskier riding patterns through their subjective belief of increased safety (see Underwood in this book). Second, the overall success of safety treatments (if not mandatory) is clearly depending on riders' willingness to use them (Beanland, et al., 2013). With respect to riders' acceptability of enhanced light configuration as conspicuity treatment, we can also report positive findings as we presented the prototypes to riders and we examined riders' acceptability and its underlying determinants (Rößger et al., 2012). Results indicated a relatively positive view of the riders concerning the light configuration compared to other conspicuity treatments (such as warning vests). The findings suggested that riders apparently disapprove conspicuity treatments when they feel that the treatment is not in line with their self-image or with the bike itself despite the fact that they perceive that the treatment would enhance their conspicuity, and, thus, their safety (in fact ratings concerning perceived effectiveness differed only slightly between the treatments). The results emphasize that often conflicting motives interfere with the use of such treatments, as riding a bike is obviously more – and strongly – associated with multiple motives compared to using other means of transportation. As results from interviews with M/C riders indicated (Turetschek et al., 2011), riding is to a great extent about an expression of life style, probably more so than driving a car. This, in turn, hints at the challenging task for designers and suppliers to create conspicuity treatments and thereby gratify, also, various aspects in addition to safety.

References

ACEM. (2004). MAIDS: In-depth Investigation of Accidents involving Powered Two-Wheelers. Report of the Association of European Motorcycle

Manufacturers. Retrieved Mai 02, 2010. Retrieved from: http://www.maids-study.eu

Baures, R., Oberfeld, D. and Hecht, H. (2010). Judging the contact-times of multiple objects: Evidence for asymmetric interference. *Acta Psychologica*, 132, 363–71.

Beanland, V., Lenne, M., Fuessl, E., Oberlader, M., Joshi, S., Bellet, T., Banet, A., Rößger, L., Leden, L., Spyropoulou, I., Yannis, G., Roebroeck, H., Carvalhais, J. and Underwood, G. (2013). Acceptability of rider assistive systems for powered two-wheelers. *Transportation Research Part F*, 19, 63–76.

Castro, C., Martinez, C., Tornay, F., Fernandez, P. and Martos, F. (2005). Vehicle distance estimations in nighttime driving: A real-setting study. *Transportation Research Part F*, 8, 31–45.

Cavallo, V. and Pinto, M. (2012). Are car daytime running lights detrimental to motorcycle conspicuity? *Accident Analysis & Prevention*, 49, 78–85.

De Graen, S., Doumen, M. and van Norden, Y. (2011). The roles of motorcyclists and car drivers in conspicuity-related motorcycle chrashes. Report-No. R-2011–25. Leidschendam: Stichting Wetenschappelijk Onderzoek Verkeersveiligheid SWOV.

DeLucia, P. and Warren, R. (1994). Pictorial and motion-based depth information during active control of self-motion: Size-arrival effect on collision avoidance. *Journal of Experimental Psychology: Human Perception and Performance*, 20, 783–9.

Franconeri, S., Hollingworth, A. and Simons, D.J. (2005). Do new objects capture attention. *Psychological Science*, 16(4), 275–81.

Gould, M., Poulter, D.R., Helman, S. and Wann, J. (2012b). Judgements of approach speed for motorcycles across different lighting levels and the effect on an improved tri-headlight configuration. *Accident Analysis & Prevention*, 48, 341–5.

Gould, M., Poulter, D.R., Helman, S. and Wann, J.P. (2012). Errors in judging the approach rate of motorcycles in nighttime conditions and the effect of an improved ligthing configuration. *Accident Analysis & Prevention*, 45, 432–7.

Hecht, H. and Savelsbergh, G. (2004). *Time-to-Contact*. Amsterdam: Elsevier.

Hole, G. (2007). *The Psychology of Driving*. Mahwah, New Jersey: Lawrence Erlbaum Associates.

Hole, G. and Tyrrell, L. (1995). The influence of perceptual set on the detection of motorcyclists using headlights. *Ergonomics*, 38(7), 1326–41.

Horswill, M., Helman, S., Ardilles, P. and Wann, J. (2005). Motorcycle Accident Risk Could be inflated by a time to arrival illusion. *Optometry and Vision Science*, 82, 740–46.

Itti, L. and Koch, C. (2000). A saliency-based search mechanism for overt and covert shifts of visual attention. *Vision Research*, 40, 1489–506.

Koornstra, M., Bijleveld, F. and Hagenzieker, M. (1997). The safety effects of daytime running lights: a perspective on daytime running lights (DRL) in the EU. Netherlands: SWOV, Institute for Road Safety Research.

McCarthy, M., Walter, L., Hutchins, G., Tong, R. and Keigan, M. (2007). Comparative analysis of motorcycle accident data from OTS and MAIDS. Wokingham, UK: Published Report PPR 168. TRL Limited.

Oberfeld, D. and Hecht, H. (2008). Effects of a Moving Distractor Object on Time-to-Contact Judgements. *Journal of Experimental Psychology: Human Perception and Performance*, 34(3), 605–23.

Rößger, L., Hagen, K., Krzywinski, J. and Schlag, B. (2012). Recognisability of different configurations of front lights on motorcycles. *Accident Analysis & Prevention*, 44, 82–7.

Rößger, L., Mühlbauer, F., Krzywinski, J. and Schlag, B. (2012). An investigation of powered-two-wheelers' acceptability towards conspicuity treatments and influencing factors. Abstract Book of the 5th International Conference on Traffic and Transportation Psychology, (p. 46). Groningen, Netherlands.

Roge, J., Douissembekov, E. and Vienne, F. (2012). Low conspicuity of motorcycles for car drivers: dominat role of bottom-up control of visual attention or deficit of top-down control? *Human Factors*, 54(1), 14–25.

Rumar, K. (2003). *Functional Requirements for Daytime Running Lights*. Ann Arbor, MI: Transportation Research Institute.

Thomson, G. (1980). The role frontal motorcycle conspicuity has in road accidents. *Accident Analysis & Prevention*, 12, 165–78.

Tsutsumi, Y. and Maruyama, K. (2008). Long Lighting System for enhanced conspicuity of motorcycles. Paper Number 07–0182, Japan. Retrieved: 15 July 2009, http://www-nrd.nhtsa.dot.gov/pdf/esv/esv20/07–0182-O.pdf

Turetschek, C., Füssl, E., Oberlader, M., Mallek, K. and Schaner, P. (2011). Interaction processes of motorcycle riders with other road users. Del. 17 of 2-BE-SAFE Project, GA No. 218703.

Underwood, G., Humphrey, K. and van Loon, E. (2011). Decisions about objects in real-world scenes are influenced by visual saliency befor and during their inspection. *Vision Research*, 51, 2031–8.

Williams, M. and Hoffmann, E. (1979). Conspicuity of motorcycles. *Human Factors*, 21, 619–26.

Wulf, G., Hancock, P. and Rahimi, M. (1989). Motorcycle conspicuity: An evaluation and synthesis of influential factors. *Journal of Safety Research*, 20, 153–76.

Chapter 7

Visual Factors Affecting Motorcycle Conspicuity: Effects of Car Daytime-running Lights and Motorcycle Headlight Design

Viola Cavallo and Maria Pinto

Visual Conspicuity of Motorcycles as a Safety Issue

Motorcycles are known to be less salient than cars, and 'look but fail to see' errors made by automobile drivers have often been mentioned to explain accidents of these vulnerable road users (Crundall et al., 2008; van Elslande and Jaffard, 2010).

Visual conspicuity or saliency is understood as the property of an object to attract attention and to be easily detected (Connors, 1975). It refers to the dissimilarity between the visual properties of the object and its local background. Such visual properties are size, brightness, colour, outline and motion (Engel, 1977; Itti and Koch, 2001; Toet and Bijl, 2003). An object is not conspicuous *per se*, but the surroundings also have to be considered as this determines contrast and may contain distractors.

The motorcycles' visual conspicuity is reduced, as compared to cars, especially because of their small size, but also because of their irregular contours. Safety measures for increasing motorcycles' visual conspicuity have been considered of crucial importance to compensate for their inherently lower detectability. The compulsory use of daytime-running lights (DRL) by motorcycles, introduced in many countries in the 1970s already, has been proven to enhance their attention-attracting features and to reduce accidents that occur because of detection failures by other vehicle drivers (Dahlstedt, 1986; Hendtlass, 1992; Muller, 1984; Olson et al., 1981; Thomson, 1980; Williams and Hoffmann, 1979; Zador, 1985). This improvement is obtained by affording a high contrast between the motorcycle headlight and the background on the one hand, and by providing motorcycles with consistent visual characteristics that facilitate their search and identification. Other conspicuity treatments such as brightly coloured motorcycles and helmets (Comelli et al., 2008) or fluorescent vests and helmet covers (Olson et al., 1981) have also been found to improve motorcyclists' conspicuity, but their safety benefit is limited as it depends on changing background characteristics (Gershon and Shinar, 2012). These conspicuity treatments have been shown to be less powerful means than motorcycle headlights lit at daytime (Olson et al., 1981).

The conspicuity benefit of motorcycles as the only vehicles with DRL is presently getting lost due to the growing practice among car drivers of turning on their lights during the day, now authorized or required in most countries. Car DRLs are likely to constitute visual distractors in the vicinity of motorcycles that affect their detection. Furthermore, motorcycles are no longer the only vehicles that use headlights in daytime and thus lose their specific visual signature. Many researchers and road-safety specialists, as well as motorcycling associations, have suspected a detrimental impact of car DRLs on motorcycle safety (Brendicke et al., 1994; Cobb, 1992; FEMA, 2006; Hörberg and Rumar, 1979; Knight et al., 2006). An adverse effect could also arise for bicyclists and pedestrians: if cars with daytime lights are more conspicuous, then unequipped users could also become less conspicuous and thus less noticeable (for example, Brouwer et al., 2004; Hörberg and Rumar, 1979).

The potential impact of car DRLs on road accidents in general, and on vulnerable road users specifically, has long been a topic of debate among road-safety specialists and has given rise to a large body of accident studies (for a review, see for example Cairney and Styles, 2003). While a large number of accident studies suggest that car DRLs improve safety (for example, Andersson and Nilsson, 1981; Koornstra et al., 1997; Tofflemire and Whitehead, 1997), Elvik (1996) has pointed out that the safety gain, generally estimated at 5 per cent to 10 per cent, depends on the surrounding luminosity, making the gain greater in Finland, for example, than in Israel. Several other studies have found little or no significant improvements (Farmer and Williams, 2002; Theeuwes and Riemersma, 1995; Wang, 2008). Some authors have drawn attention to the possible negative effects of car DRLs on unprotected users (for example, Theeuwes and Riemersma, 1995; Wang, 2008). Others recommend using dedicated low-intensity car DRLs (rather than standard passing beam headlights) to limit adverse effects on motorcycles (for example, Knight et al., 2006), and still others advise implementing measures that increase the conspicuity of motorcycles (for example, Rumar, 2003). However, most studies have not found any detrimental effects of car DRLs (for example, Cobb, 1992; NHTSA, 2000; Paine, 2003; Riemersma et al., 1987; Rumar, 2003; Schönebeck et al., 2005).

A surprisingly low number of experimental studies have addressed the question of whether car DRLs affect motorcycle detection, and more generally whether they have an effect on the conspicuity of vulnerable road users. Regarding possible masking effects, Cobb (1999) studied various car-DRL intensity levels in real-world conditions. He found that DRLs improved the detection of automobiles in cloudy weather without decreasing the detection of motorcycles and bicycles, provided no high-intensity lamps (greater than 600 cd) were used. Brouwer et al. (2004) used slides of traffic scenes including one car and sometimes another road user. In these very poor visual conditions they even observed a slight detection advantage for the other road user when car DRLs were on. Brendicke et al. (1994) also used slides of traffic scenes and observed a detrimental effect of car DRLs on motorcycle detection, but the ecological validity of their findings has often been questioned insofar as the participants' task was to count the number of vehicles in the scene.

In this context, it seemed critical to determine whether the use of daytime lights by automobilists alters motorcycle conspicuity, and if so, to find new ways of improving it.

Methodological Considerations

When considering previous studies it is not clear whether car DRLs affect the automobilists' detection and recognition of motorcycles and other vulnerable road users. It can be assumed that the inconclusive results are due to different methodological choices, in particular to how the very notion of visual conspicuity was operationalized. Most studies have used simplified experimental tasks and situations, such as searching for a motorcycle in an impoverished environment and/ or with an unlimited or long exploration time. Such situations are not representative of the attentional demands of real driving, however, and are therefore not very likely to generate perception errors. We contend that the visual conspicuity of motorcycles should be assessed in complex environments, with a time-limited task where the observer has to 'notice' a motorcycle, not to purposely look for one. Gershon and Shinar (2012) have shown that fully alerted drivers almost perfectly detect motorcycles, but that at normal levels of expectancy the motorcycle detection rates are considerably lower. In short, the experimental situation must call upon selective attention, in such a way that attentional conspicuity rather than search conspicuity is at stake (Hughes and Cole, 1984).

The present study comprised two experiments that investigated various visual factors likely to affect motorcycle conspicuity. These factors pertain to potential distractors in the motorcycle's visual environment on the one hand, namely car DRLs, and to the motorcycle's visual features, such as their front lights, on the other hand.

The first experiment investigated the effects of automobile DRLs on motorcycle detection and recognition. The aim was to determine the conditions under which car lights are likely to create 'visual noise' that hampers the perception of motorcycles and other vulnerable road users. The participants were briefly presented with real-world photographs and had to indicate whether they had seen vulnerable users – motorcyclists, bicyclists and pedestrians, or none of them. A short presentation time of 250 ms was chosen to simulate an information intake corresponding to just one glance (Crundall et al., 2008). A complex urban environment and low-luminosity conditions (overcast skies) were used. The photos represented a great variety of complex urban traffic situations and contained a good number of details, which would have been impossible to obtain with computer-generated images. As target distance (and subsequent target size) and eccentricity are known to influence object detection (Engel, 1971; Jenkins and Cole, 1986), these factors were varied. The choice of three visual targets appearing in a complex environment and at a great number of locations contributed to minimize the building of specific expectancy-driven search strategies. We hypothesized that in these conditions, the use of daytime lights by car drivers decreases the conspicuity of motorcycles, bicycles and pedestrians.

The second experiment aimed to assess various innovative motorcycle headlight configurations to identify light arrangements than can enhance motorcycle conspicuity in a car-DRL environment, by giving them a sort of new and unique visual signature.

Are Daytime-running Lights Visual Distractors that Hamper Motorcycle Conspicuity?

To answer this question, a set of colour photographs of traffic at urban intersections was prepared. The photos were edited in Photoshop so as to unify contrast levels, eliminate undesirable elements (for example, billboards), and include distractors (car DRLs) and visual targets (motorcyclists, cyclists and pedestrians) at different distances and eccentricities (see Figure 7.1). Among the 180 pictures presented, half of them contained a visual target. The pictures were displayed on a 40" LCD screen with high quality visual rendering.

Twenty four participants (M=36.6 years old) took part in the experiment. All of them were licensed drivers with normal vision and normal visio-attentional performances. The rates of correct detections of the three kinds of vulnerable road users were determined.

The results show that vulnerable road users were significantly more often detected when car DRLs were off (in 56 per cent of cases) than when they were on (in 49 per cent of cases). When looking more specifically on motorcycle detection (Figure 7.2), a difference of 6 percentage points (60 per cent vs. 66 per cent) was found according to the presence or absence of car lights in their vicinity. Regarding the bicyclists and pedestrians – the two types of road users that did not have lights in our experiment – it could be expected that the masking effect of car DRLs would be even more detrimental for these users, but our results did not support this assumption (Figure 7.2). To the same extent as observed for motorcycles, bicyclists and pedestrians were more difficult to perceive when surrounded by cars with their DRLs on: the detection rates dropped by 9 (49 per cent vs. 40 per cent) and 5 percentage points (53 per cent vs. 48 per cent), respectively. As a whole the adverse effect of DRLs was of comparable magnitude for the three types of vulnerable road users.

Apart from the systematic detrimental effect of car DRLs on all vulnerable road users, it was observed that motorcycles were better detected (63 per cent) than pedestrians (51 per cent) and bicycles (45 per cent) (see Figure 7.2). This was probably due to their greater conspicuity owing to their slightly bigger size and especially to the presence of their headlights that were easy to see. Pedestrians, in turn, were more frequently detected than cyclists. However, the differences in detection rates between the various road users should be taken with caution because in real-world conditions the relatively fast (arm and/or leg) movements made by

**Figure 7.1   Examples of photographs showing motorcycles in the near/
central condition and in a car DRL-off (upper) or car DRL-on
(lower) environment**

pedestrians and cyclists are likely to provide motion cues that make them more
conspicuous than in the present experiment where stationary pictures were used.

The experiment also replicated previous findings on the influence of target
distance and eccentricity on detection performance: the detection rate increased
by 39 and 32 percentage points, respectively, when the vulnerable road users
appeared at short distances and/or in the centre of the picture. Regarding the effect
of distance, this variable determined the object's angular size, which has been
shown to be an important conspicuity-influencing feature (for example, Engel,

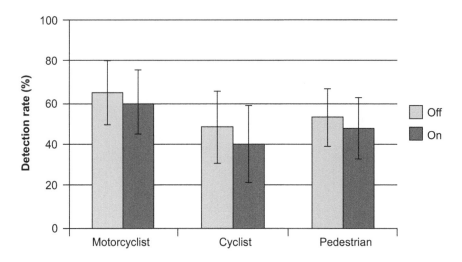

Figure 7.2   Mean and standard deviation of detection rate (%) for motorcyclists, cyclists and pedestrians, according to whether car DRLs were off or on

1971; Jenkins and Cole, 1986). Eccentricity is also known to affect the detection of objects in a visual scene (for example, Carrasco and Chang, 1995; Engel, 1977). In our experiment, the participants had to initially fixate the centre of the screen, so no (or only small) eye movements were required to see and identify the targets when the photograph was displayed. By contrast, detecting off-centre targets needed larger eye movements and more time.

Target distance and eccentricity were found to influence the extent to which car DRLs affected motorcycle detection. The car-DRL effect was particularly strong when the motorcycle was far away (detection rate differences of 14 per cent), and when it was located in the centre of the visual scene (detection rate differences of 9 per cent). The negative effect of car DRLs at greater distances suggests that the illuminated car lamps generated competing light patterns especially in conditions where the motorcycles were hard to see because their greater distance from the observer and their smaller angular size. At shorter distances, where the angular size of the motorcycle and thus its inherent conspicuity were greater, car DRLs had little or no impact. More surprisingly, the adverse effect of car DRLs was observed even when the motorcycles were located in the central part of the observers' visual field, which is a favourable condition for target detection. It would seem that in these conditions, it was specifically motorcycle identification that was impeded due to a mix-up of light sources between cars and motorcycles.

All in all, these findings unequivocally indicate that car DRLs are visual distractors that hinder the perception of vulnerable road users. With respect to motorcycles, car lights illuminated at daytime seem to constitute light patterns that compete with the motorcycle's headlights, and thus make their detection and recognition more difficult, notably when motorcycles are at far distance and in the centre of the visual scene.

The present findings are not in line with most of the earlier experimental studies (for example, Brouwer et al., 2004). Several reasons can be brought to bear to explain these discrepancies. The use of photographs of real traffic scenes in an urban context allowed us to create realistic, complex and unpredictable visual conditions in which a road user's conspicuity was likely to be critical. Furthermore, the brevity of scene exposure in our study enabled us to simulate situations of rapid information intake where automobilists had only enough time to glance before making a decision. Although this strategy is certainly not the prevailing one, it is probably used in some cases (when the driver is in a hurry, for example) and is bound to generate the kind of perceptual errors that cause accidents.

While we can reasonably think that the present study was effective in recreating relevant real-world task demands when car drivers interact with motorcycles, how can we then explain that the detrimental effect of car DRLs observed here in our experiment has scarcely been found in accident analyses? Several authors suggest that car DRLs can be beneficial for vulnerable road users because they can see cars better and avoid collisions with them (Andersson and Nilsson, 1981; Rumar 2003). In this way, the increased conspicuity of automobiles for vulnerable road users could compensate for the decreased conspicuity of vulnerable road users for car drivers, so that detrimental effects of car DRLs do not show up in global accident statistics. The findings of our study indicate, however, that the safety of vulnerable road users could be further improved if car drivers better detect these users in a car-DRL environment. We have in particular studied this question with regard to motorcycle conspicuity.

Can Motorcycle Conspicuity Be Improved by Innovative Headlight Design?

In order to counteract the detrimental effect of car DRLs on motorcycle detection and recognition, it seemed essential to seek new means of providing them again with a unique visual signature, which clearly differentiates them from cars and facilitates their perception by the other road users. Some recent studies have already identified innovative motorcycle headlight configurations likely to constitute a new visual signature for these users and helping to better recognize them on the road. Rößger et al. (2012) recommend a T-shaped light configuration that was shown to be more quickly identified by the other road users than a standard configuration. Gershon and Shinar (2013) provided evidence for the effectiveness of an alternating-blinking lights system producing a phi-phenomenon and mounted on the rider's helmet.

The present study explored alternative solutions that seemed easier to implement on motorcycles. It evaluated three visual configurations that focus on the colour of the motorcycles' frontal headlights as well as on their positioning and layout.

The colour of the frontal headlight could be a means of easily differentiating motorcycles from other lit road users. As the human eye is very sensitive to yellow or similar hues, this most attention-getting colour in a white car-DRL environment was considered to provide a highly distinctive feature. We made the assumption that the particular human sensitivity to yellow combined with the low prevalence of this lighting colour in traffic can enhance motorcycle conspicuity.

Concerning the positioning and layout of headlights, two configurations were chosen. The first one consisted of a standard frontal headlight plus two additional lights on the rearview mirrors. It is based on the assumption of a dedicated face recognition process (for example, Chellappa et al., 1995). This triangle arrangement is thought to be reminiscent of a 'face', which is naturally highly recognised by people and could thus enhance motorcycle detection, as demonstrated by Maruyama et al. (2009) in night-time visual conditions. The second arrangement consisted of adding a light on the motorcyclist's helmet to accentuate the vertical dimension of the motorcycle, which could favour a better detection, but also a better perception of its distance, speed and time to arrival (Tsutsumi and Maruyama, 2008).

The methodology used was very similar to the one of the first experiment. The only differences pertain to the fact that merely conditions with car-DRLs lit were used, and that four groups of 15 participants each (M=28 years) took part in the experiment. The groups were matched with respect to age, gender, as well as driving and riding experience. Each group was submitted to one of the four motorcycle headlight configurations: standard (one frontal white headlight),

a)   b)   c)   d)

Figure 7.3   Examples of the four motorcycle lighting configurations:
a) standard; b) yellow; c) helmet; and d) triangle

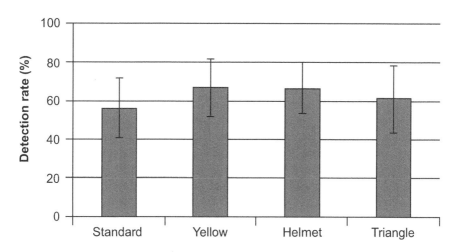

Figure 7.4   **Mean and standard deviation of detection rate (%) for motorcycles equipped with standard, yellow, helmet and triangle configurations**

yellow (one frontal yellow headlight), triangle (standard plus two position lights on rear view mirrors) and helmet (standard plus one position light on the helmet) (Figure 7.3). All other experimental variables, material, task and procedure were the same as before.

Motorcycles were best detected when they were equipped with a yellow headlight (67 per cent) (Figure 7.4). This configuration gave rise to an 11 per cent higher detection rate than the standard configuration (56 per cent). When considering interactions with distance and eccentricity, the yellow headlight was significantly more effective than the standard configuration when the motorcycle was at far distances (54 per cent vs. 39 per cent) and at central positions in the visual scene (90 per cent vs. 71 per cent). No significant differences were found between the yellow and the standard configurations at near distance and in off-centre positions in the visual scene. The colour coding seemed to provide the most effective means of increasing dissimilarity between motorcycles and cars and to successfully compensate for the detrimental effect of car DRLs on motorcycle conspicuity. These findings are in line with studies that highlight the importance of colour coding in visual search and show its advantage over size-based and shape-based coding (Christ, 1975; Nowell, 1997).

The helmet configuration (67 per cent) was also globally better detected than the standard configuration (56 per cent) (Figure 7.4). When looking at distance and eccentricity conditions, we noticed significantly better detection performances for motorcycles at far distances when they were equipped with the helmet configuration (61 per cent) as compared with the standard one (39 per cent). The reason for the effectiveness of the helmet configuration is probably related to its height: it represents the highest source of light in the visual scene and thus forms a new kind of visual signature that distinguishes motorcycles from other lighted road users. It specifically facilitates motorcycle perception when it is difficult to see due to its far distance and small size, whereas it does not provide a significant conspicuity advantage at near distances, nor in central or off-centre positions in the visual scene.

The triangle configuration (61 per cent) tended to be better detected than the standard configuration (56 per cent) (Figure 7.4), but the five-point difference was not statistically significant. Although it contained easily recognizable elements (a kind of face) and constituted an original light arrangement with regard to existing vehicle light configurations, this light arrangement proved to lack features than could clearly distinguish it from car DRLs, probably because position lights on rear mirrors remain in the lighted area of car DRLs in dense urban traffic. Similarly to motorcycles with white standard headlights whose detection was reduced in car-DRL environments, as shown before, the triangle configuration did not prevent the motorcycle from being masked by car DRLs. This result is contrary to Maruyama et al.'s (2009) study, attesting improved detection of motorcycles equipped with a triangle configuration, in night-time (high-contrast) conditions and in a quite simplified (computer-generated) visual environment. It can be assumed that the triangle configuration enhanced motorcycle detection in quite poor visual conditions, but in the situation where motorcycles were embedded in a car-DRL environment, they have probably been masked by the other light sources. Furthermore, in the daytime conditions of our experiment, the cluttered urban background may have constituted additional distracting features that made the triangle configuration less discernable.

Conclusions

Our study has identified several visual factors that influence motorcycle visual conspicuity in a complex environment and under time constraint, conditions that are likely to produce perceptual errors when automobile drivers interact with motorcyclists. The detrimental effect of car DRLs in the vicinity of motorcycles, acting as visual distractors, has been clearly demonstrated.

While both the yellow and the helmet configurations have been shown to be effective in improving motorcycle visual conspicuity, the colour coding, by its highly distinctive features, seems to have the greatest potential for offsetting the detrimental effects of car DRLs. Both configurations represent quite simple and ergonomically realistic solutions.

It is interesting to note that the yellow and the helmet configurations enhanced motorcycle detection especially in conditions where the car-DRL environment was shown to have major detrimental effect, as seen in the first experiment, that is, when motorcycles were far away and in the central position of the visual scene. These findings confirm that the car-DRL effects observed were actually due to competing light patterns, whose adverse effects could be reduced by the conspicuity enhancements produced by the yellow and helmet configurations. In other words, when it was difficult to see a motorcycle because of its small angular size at great distance in a car-DRL environment, adding a visual signature like a yellow headlight or a helmet light enhanced the motorcycle's conspicuity and improved its detection. In the central part of the visual scene, where motorcycles were easily detected, we can assume that the yellow light acted as a conspicuity feature that improved more specifically motorcycle identification.

With regard to the validity of the present study, our findings need to be confirmed by calling on complementary methods, namely by including motion information (as present in dynamic visual scenes) and, probably still more important, by using higher contrast and luminosity levels closer to real-world conditions.

Despite these limitations, our study draws attention to safety issues for motorcycles – and also for other vulnerable road users – in road environments that contain more and more competing vehicle light sources. Research should be intensified to improve their conspicuity and make their travel safer. Regarding motorcycles, we believe that our study indicates promising ways for defining headlight configurations that can provide these users with a new visual signature.

References

Andersson, K. and Nilsson, G. (1981). *The Effects on Accidents of Compulsory Use of Running Lights During Daylight in Sweden.* VTI, Linköping, Sweden.

Brendicke, R., Forke, E. and Schäfer, D. (1994). Auswirkungen einer allgemeinen Tageslichtpflicht auf die Sicherheit motorisierter Zweiräder. In: 6. Fachtagung 'Motorrad', VDI Berichte 1159, Düsseldorf, Germany, 283–318.

Brouwer, R.F.T., Jansen, W.H., Theeuwes, J., Duistermaat, M. and Alferdinck, J.W.A.M. (2004). *Do Other Road Users Suffer from the Presence of Cars that Have Their Daytime Running Lights On?* TNO, Soesterberg, The Netherlands.

Cairney, P. and Styles, T. (2003). Review of the Literature on Daytime Running Lights (DRL). Australian Transport Safety Bureau, Canberra, Australia.

Carrasco, M. and Chang, I. (1995). The interaction of objective and subjective organizations in a localization search task. *Perception and Psychophysics*, 57, 1134–50.

Christ, R.E. (1975). Review and analysis of color coding research for visual displays. *Human Factors*, 17, 542–70.

Cobb, J. (1992). *Daytime Conspicuity Lights.* Transport Research Laboratory, Crowthorne, Berkshire, United Kingdom.

Cobb, J. (1999). *Vehicle Lighting and Signalling.* Transport Research Laboratory, Crowthorne, Berkshire, United Kingdom.

Comelli, M., Morandi, A., Magazzù, D., Bottazzi, M. and Marinoni, A. (2008). Brightly coloured motorcycles and brightly coloured motorcycle helmets reduce the odds of a specific category of road accidents: A case-control study. *Biomedical Statistics and Clinical Epidemiology*, 2, 1–8.

Connors, M.M. (1975). *Conspicuity of Target Lights: The Influence of Color*. NASA, Washington, DC.

Chellappa, R., Wilson, C.L. and Sirohey, S. (1995). Human and machine recognition of faces: A survey. *Proceedings of the IEEE Conference 83*, 5, 705–40.

Crundall, D., Humphrey, K. and Clarke, D. (2008). Perception and appraisal of approaching motorcycles at junctions. *Transportation Research Part F: Traffic Psychology and Behaviour*, 11, 159–67.

Dahlstedt, S. (1986). A Comparison of Some Daylight Motorcycle Visibility Treatments. VTI, Linköping, Sweden.

Elvik, R. (1996). A meta-analysis of studies concerning the safety effects of daytime running lights on cars. *Accident Analysis and Prevention*, 28, 685–94.

Engel, F.L. (1971). Visual conspicuity, directed attention and retinal locus. *Vision Research*, 11, 563–75.

Engel, F.L. (1977). Visual conspicuity, visual search and fixation tendencies of the eye. *Vision Research*, 17, 95–108.

Farmer, C.M. and Williams, A.F. (2002). Effects of daytime running lights on multiple-vehicle daylight crashes in the United States. *Accident Analysis and Prevention*, 34, 197–203.

FEMA (2006). Saving [car drivers'] lives with daytime running lights: Consultation paper. FEMA's comments. EC Web site, Brussels, Belgium.

Gershon, P., Ben-Asher, N. and Shinar, D. (2012). Attention and search conspicuity of motorcycles as a function of their visual context. *Accident Analysis and Prevention*, 44, 97–103.

Gershon, P. and Shinar, D. (2013). Increasing motorcycles attention and search conspicuity by using alternating-blinking lights system (ABLS). *Accident Analysis and Prevention*, 50, 801–10.

Hendtlass, J. (1992). Appendix C: Literature review. Daytime running lights. In *Social Development Committee Inquiry into Motorcycle Safety in Victoria*. Parliament of Victoria, Melbourne, Australia.

Hörberg, U. and Rumar, K. (1979). The effects of running lights on vehicle conspicuity in daylight and twilight. *Ergonomics*, 22, 165–73.

Hughes, P.K. and Cole, B.L. (1984). Search and attention conspicuity of road traffic control devices. *Australian Road Research Board*, 14, 1–9.

Itti, L. and Koch, C. (2001). Computational modelling of visual attention. *Nature*, March, 194–203.

Jenkins, S.E. and Cole, B.L. (1986). Daytime conspicuity of road traffic control devices. *Transport Research Record*, 1093, 74–80.

Knight, I., Sexton, B., Bartlett, R., Barlow, T., Latham, S. and Mccrae, I. (2006). Daytime running lights (DRL): A review of the reports from the European Commission. TRL, Crowthorne, Berkshire, United Kingdom.

Koornstra, M., Bijleveld, F., Hagenzieker, M.P. (1997). The safety effects of daytime running lights. SWOW, Leidschendam, The Netherlands.

Maruyama, K., Tsutsumi, Y., Murata, Y. (2009). Face design, lighting system design for enhanced detection rate of motorcycles. *Journal of the Society of Automotive Engineers of Japan*, October, 27–30.

Muller, A. (1984). Daytime headlight operation and motorcyclist fatalities. *Accident Analysis and Prevention*, 16, 1–18.

NHTSA (2008). Traffic safety facts. Washington, DC, USA.

Nowell, L.T. (1997). Graphical encoding for information visualization: Using icon color, shape, and size to convey nominal and quantitative data. PhD thesis, Virginia Polytechnic Institute and State University, Blacksburg, Virginia, USA.

Olson, P.L., Halstead-Nussloch, R. and Sivak, M. (1981). The effect of improvements in motorcycle/motorcyclist conspicuity on driver behavior. *Human Factors*, 23, 237–48.

Paine, M. (2003). A review of daytime running lights. NRMA Motoring & Services and RACV, Sydney, Australia.

Riemersma, J.B.J., Wertheim, A.H., Welsh, M. and Baker, P.J., 1987. De opvallendheid van het attentielicht. Institute for Perception, TNO, Soesterberg, The Netherlands.

Rößger, L., Hagen, K., Krzywinski, J. and Schlag, B. (2012). Recognisability of different configurations of front lights on motorcycles. *Accident Analysis and Prevention*, 44, 82–87.

Rumar, K. (2003). Functional requirements for daytime running lights. University of Michigan Transportation Research Institute, Ann Arbor, MI, USA.

Schönebeck, S., Ellmers, U., Gail, J., Krautscheid, R. and Tews, R. (2005). Abschätzung möglicher Auswirkungen von Fahren mit Licht am Tag (Tagfahrleuchten/ Abblendlicht) in Deutschland. Bundesanstalt für Strassenwesen, Bergisch Gladbach, Germany.

Theeuwes, J. and Riemersma, J. (1995). Daytime running lights as a vehicle collision countermeasure: The Swedish evidence reconsidered. *Accident Analysis and Prevention*, 27, 633–42.

Thomson, G.A. (1980). The role frontal motorcycle conspicuity has in road accidents. *Accident Analysis and Prevention*, 12, 165–78.

Toet, A. and Bijl, P. (2003). Visual conspicuity. *Encyclopedia of Optical Engineering*, 2929–35.

Tofflemire, T.C. and Whitehead, P.C. (1997). An evaluation of the impact of daytime running lights on traffic safety in Canada. *Journal of Safety Research*, 28, 257–72.

Tsutsumi, Y. and Maruyama, K. (2008). Long lighting system for enhanced conspicuity of motorcycles. *JSAE Transaction*, 39, 119–24.

Van Elslande, P. and Jaffard, M. (2010). Typical human errors in traffic accidents involving powered two-wheelers. In Proceedings of the 27th International Congress of Applied Psychology, Melbourne, Australia, 1363–4.

Wang, J.-S. (2008). The effectiveness of daytime running lights for passenger vehicles. NHTSA, Washington, DC.

Williams, M.J. and Hoffmann, E.R. (1979). Motorcycle conspicuity and traffic accidents. *Accident Analysis and Prevention*, 11, 209–24.

Zador, P.L. (1985). Motorcycle headlight-use laws and fatal motorcycle crashes in the US, 1975–83. *American Journal of Public Health*, 75, 543–6.

PART III
Case Studies with Additional Focus on Top-down Influences

Chapter 8

Is the Poor Visibility of Motorcycles Related to Their Low Sensory and Cognitive Conspicuity or to the Limited Useful Visual Field of Car Drivers?

Joceline Rogé and Fabrice Vienne

Introduction

Over approximately ten years in France, the proportion of motorcyclists killed rose from 9.6 per cent in 1994 to 14.2 per cent in 2003, and from 9.4 per cent in 1994 to 13.6 per cent in 2003 for those injured (Filou et al., 2005). In Europe (data from 14 European countries), the percentage of motorcycle rider and passenger fatalities was 16.1 per cent of the total number of road accident fatalities in 2006 (Leitner et al., 2008). In the MAIDS (Motorcycle Accident In-Depth Study) study, researchers characterized the nature and causes of motorcycle crashes in order to identify several categories of factors that can contribute to a crash. Human error (for perception/attention, comprehension, decisions, behaviour) was one of these categories and the primary cause of motorcycle crashes. In accidents where human error was identified as the primary cause of the collision, 35.6 per cent were the result of a lack of attention and/or perception on the part of the car driver (compared with 11.9 per cent caused by error on the part of the motorcycle rider). One explanation that could be advanced for accident rates among motorcyclists is their poor visibility to other drivers (Williams and Hoffmann, 1979; Shinar, 2007). In this study we present four hypotheses which may explain, at least in part, two-wheeled motorized vehicles' lack of visibility for other drivers. The first of these is based on the low sensory conspicuity of two-wheeled motorized vehicles. The second hypothesis is based on the lack of cognitive conspicuity of such vehicles for other drivers. And the two remaining hypotheses postulate the existence of individual differences in drivers' visuo-attentional abilities and ocular behaviour. These hypotheses were all tested as part of the European project (the 2BeSafe Project, number 218703), some of whose results have already been published (Rogé et al., 2012).

Motorcycle Sensory Conspicuity

Sensory conspicuity refers to the extent to which an object can be distinguished from its environment, due to its physical characteristics: angular size, eccentricity in relation to the point of gaze, brightness against the background, colour and so on (Engel, 1971, 1974; Cole and Hughes, 1988; Wulf et al., 1989; Hancock et al., 1990). In other words, sensory conspicuity reflects an object's ability to attract visual attention and to be located precisely as a result of its physical properties.

Wells et al. (2004) quantify the accident risk associated with motorcycle riding by comparing the characteristics of motorcyclists involved in accidents with those of randomly selected motorcyclists (Wells et al., 2004). Those authors identified a relationship between helmet colour and accident risk. Comelli and colleagues (2008) reached a similar conclusion by analysing accidents for which the recorded cause was the low conspicuity of the motorcycle (Comelli et al., 2008). However, the contrast between the motorcycle and the background against which it appears is more important than the colour of the motorcycle itself (Watts, 1980; Hole et al., 1996). In a recent experiment carried out using a car-driving simulator, the sensory conspicuity of motorcycles riding between lanes of vehicles was improved by a modification in the level of colour contrast between the motorcycle and traffic (Rogé et al., 2010). Researchers found that a high colour contrast improves the conspicuity of motorcycles in certain speed and traffic conditions. This effect is also dependent on the age of the car drivers. It should be noted that in the experiment in question, participants had to detect motorcycles only in their rearview mirrors and in traffic. It reasonable to ask ourselves if the positive effect of a high level of colour contrast could be partly (or even mainly) due to the possibility of anticipating from where and in what circumstances (for example in a stream of vehicles) the motorcycle might appear. In order to eliminate this doubt, the task used in the 2Be-Safe project consisted of detecting a motorcycle appearing from one of several directions in the road environment and in differing traffic conditions. The level of colour contrast between the motorcycle and the background varied (that is, the grey macadam of the road surface).

We suggested that a significant contrast between the colour of the road surface and the colour of the motorcycle increases the visibility of the motorcycle for car drivers (*sensory conspicuity hypothesis*).

Motorcycle Cognitive Conspicuity

According to Theeuwes, attentional focusing is not only exogenously controlled (that is, automatically determined by the physical properties of the environment) but could also be influenced by the demands of the task (Theeuwes, 1991b). This means that a highly salient object does not automatically attract attention to its location. Visual selection is considered to be endogenously controlled when the search is mediated by expectations about the properties of task-relevant objects such as their location, colour, shape, luminance or size. Endogenous control is most clearly in operation when attention is directed to where a target is likely to be found. When a search objective is given, a scene schema based on global features is activated and provides a search path to those locations in which, in view of the scene's prototypical representations, the object of the search is likely to be found. Theeuwes (1991b) believes that the momentary need for information could play a key role in the process of actively directing a driver's attention. The driver would be able to engage in active filtering based on knowledge related to the nature of probable stimulus inputs. It has been suggested that these top-down processes explain, at least in part, the low conspicuity of motorcycles. Cognitive conspicuity is linked to the fact that an observer's focus of attention is strongly influenced by his/her expectations, objectives and knowledge.

Cognitive conspicuity therefore highlights top-down processes. According to Hole et al. (1996), in many cases, inappropriate expectations may be more influential in accident causation than the motorcyclist's physical properties. It is possible that some car drivers have inappropriate expectations about what is likely to happen next based on their previous experience because of infrequent exposure to a situation (Hurt et al., 1981; Wulf et al., 1989; Rumar, 1990; Hole and Tyrrell, 1995; Langham et al., 2002). It is also possible that car drivers misinterpret what they see or where the motorcycle might appear from (Wulf et al., 1989; Brown, 2002; Langham et al., 2002; Horswill et al., 2005). The corollary of these suggestions is that car drivers who also ride a motorcycle (motorcyclist-motorists) could therefore use their riding knowledge when driving a car and this may help them to detect and avoid collisions with motorcycles. Two observations support this reasoning: a high proportion of car drivers involved in collisions with motorcyclists do not have a licence to ride a motorcycle, and car drivers who also hold a motorcycle licence have been shown to be responsible for fewer motorcycle-car collisions than drivers who do not have one (Hurt et al., 1981; Wulf et al., 1989; Magazzù et al., 2006; Comelli et al., 2008).

In the 2BeSafe project, we suggested that car drivers who also have a motorcycle licence and who ride regularly would detect motorcycles more easily because of their riding knowledge (*cognitive conspicuity hypothesis*).

Useful Visual Field and Ocular Behaviour of Car Drivers

The useful visual field was first defined by Mackworth as the area around the fixation point in which information is briefly stored and interpreted during a visual task (Mackworth, 1965). In other words, this field corresponds to the part of the peripheral visual field around the fixation point inside which sources of information can be processed at a single glance, that is, without any movement of the eyes or the head (Sanders, 1970; Scialfa et al., 1987; Ball et al., 1988). It is generally estimated by instructing the participant to carry out a dual task, one involving signals in the central part of the visual field and the other one involving signals in the peripheral part of the visual field. The useful visual field is assessed on the basis of performance in the peripheral task in which the participant has to detect the presence of a signal located at different eccentricities in his or her visual field. Due to the attentional workload related to the central task, the useful visual field is smaller than the peripheral visual field. Some researchers have also called this field the 'functional visual field' or 'conspicuity visual field' (Engel, 1974; Ikeda et al., 1979; Williams, 1982; Wood et al., 1993).

Assessment of the useful visual field during driving has led to the conclusion that the size of this field is not constant (Rogé et al., 2002, 2003, 2004 and 2008; Rogé and Pébayle, 2009). Indeed, a driver's useful visual field deteriorates in relation to factors specific to the driving task, such as duration of driving and speed. The field can vary during the driving task and its size can differ in relation to features specific to the driver, such as his/her state of alertness and age. If the driver's useful visual field is reduced, he/she is slower to detect information, such as obstacles or other vehicles, on the road (Rogé et al., 2005), exposing himself (or herself) and other road users to a potential accident.

In a previous experiment carried out using a driving simulator, we observed an interesting result concerning the ability of young car drivers to detect a motorcycle on the road because of their riding knowledge. Participants had to distinguish a change in colour of a signal on a vehicle they were following. They had to simultaneously detect a static motorcycle which briefly appeared on the road either in front of them or behind them (at 10°, 20°, 30° and 44°, the last position refers to a motorcycle riding in the left lane behind the participant and appearing in the left-side mirror). Their useful visual field had been evaluated prior the experiment in a test carried out on a computer. Drivers with a limited useful visual field were those who detected the fewest motorcycles on the road. The difference in performance between the two groups (drivers with a limited useful visual field and drivers with an extended one) was essentially due to their capacity to detect the motorcycle when it was behind their vehicle and appearing in the rearview mirror. Indeed, car drivers with an extended field detected 53.4 per cent more motorcycles in this condition than the other group. We put forward the hypothesis that motorists who also have a motorcycle licence and who ride regularly detect motorcycles more easily. The difference in performance between the two groups could also be explained by the specific aptitudes developed by motorcyclist-motorists when riding their motorcycle in order to guarantee their own safety. The motorcyclist-motorists may collect and/or process visual information from the environment (irrespective of the type of information) more quickly than non-motorcyclist-motorists when they are engaged in a task which requires them to split their attention across the whole of their useful visual field. It is also possible that motorcyclist-motorists develop an extended useful visual field, which enables them to detect elements appearing in the road environment more rapidly (*useful visual field hypothesis*).

If it is the case that motorcyclists develop certain specific aptitudes (such as a larger useful visual field and/or an ability to rapidly process visual information in this field), these aptitudes may have an effect on their visual search patterns when driving a car. Motorcyclists and car drivers have been shown to differ in their visual search patterns. According to Nagayama et al., motorcyclists look more frequently at the road surface immediately ahead (than do car drivers), and their visual field encompasses a larger proportion of road surface in order to detect any surface irregularities or other road-surface-based hazards (Nagayama et al., 1980). Moreover, mean fixation duration is longer when driving a car than when riding a motorcycle. Tofield and Wann found that motorcyclists who were driving a car looked further ahead than car drivers who had no motorcycle riding experience (Tofield and Wann, 2001). It is therefore possible that motorcyclists, when driving a car, use special visual strategies developed during motorcycle riding and which could be linked to the characteristic of their useful visual field. Motorcyclist-motorists may use specific visual strategies compared to non-motorcyclist-motorists (*ocular behaviour hypothesis*).

Method

Participants

The sample consisted of 42 car drivers (32 years old, SD=5.2), 21 motorcyclist-motorists and 21 non-motorcyclist-motorists whose visual acuity was greater than or equal to 8/10. All participants had been in possession of a driving licence for light vehicles for 14.1 years (SD=5.4) and the motorcyclist-motorists had also held a motorcycle licence for over 9 years (SD=6.5). This group of participants rode their motorcycles regularly (4.8 times per week, SD=6.8).

The test took place in a car-driving simulator (in the LEPSIS unit at IFSTTAR), which enabled us to reproduce the same driving conditions for all participants, and ensured absolute safety for the participants. This fixed-base simulator includes a cabin, four computers and four screens, onto which the simulated images are projected. Three of the screens (angle of vision: 150° X 40°) were used for the road scene ahead, and one large screen was used for the rear view. Participants were able to see the rear view using two flat rearview mirrors: the offside wing mirror and the inside rearview mirror. The Face LAB 4 system (from Seeing Machines Ltd) was used to measure ocular behaviour in real time during the test. This system records data relating to the position and inclination of the participant's head in 3D, and glance direction.

Procedure

Measurement of participants' useful visual field was carried out before studying their perception of motorcyclists. The size of the useful visual field was estimated during driving using the same protocol as in the study by Rogé et al. (Rogé et al., 2002, 2003, 2004, 2008; Rogé and Pébayle, 2009). Participants had to detect a change in colour of a disc which appeared briefly and intermittently on the rear window of the vehicle they were following. They also had to detect 48 peripheral signals (red dots) appearing briefly at several points of eccentricity on the road (from 4° to 12°). The test lasted 15 minutes. Participants had to respond quickly and without making any mistakes. The percentage of peripheral signals detected in the road environment defined the extent of their useful visual field. The reaction time for detecting peripheral signals was used to evaluate the speed at which participants processed the visual information available in their useful visual field.

The Face LAB system was then adjusted for the test (calibration of head and eye monitoring, calibration of the screen and verification of tracking quality).

The driving test then began. The test comprised three sessions of approximately 12 minutes (SD=0.86) with a break between each session to prevent phenomena such as fatigue or simulator sickness from affecting driving performance. During the three sessions, participants drove on secondary roads (where the speed limit was 90 km/h) and on a motorway (where the speed limit was 130 km/h). They also crossed junctions and roundabouts (where the speed limits were 50 km/h and 30 km/h respectively). The road environment included houses, trees, wooded areas, signposts and advertising hoardings in order to orient the attention of the participants to the whole road scene and to distract them from the motorcycles they had to detect, as might be the case when driving in a real environment. The scenarios were created in order to simulate a 'natural' driving task, as far as possible when using a car-driving simulator. Traffic therefore included small cars, vans, buses, lorries and motorcycles. The important point is that participants were unable to anticipate when and from where a motorcycle might appear because they never came back to the same section of the circuit and they had to detect motorcycles in several different situations. Their instructions were to go to a specific town (following a vehicle in the first scenario and using the road signs in the next two scenarios) and to flash their headlights as soon as they detected a motorcycle on the road. Participants were also required to adhere to the Highway Code.

Throughout the experiment, participants had to detect motorcycles which differed according to their level of colour contrast. In order to vary their sensory conspicuity, the colour of the body of the motorcycle, the clothes and the helmet of the motorcyclist were modified in order to obtain a high level of colour contrast (the red motorcycle) or a low one (the grey motorcycle) with the background (the grey macadam of the road surface). The order of the 16 motorcycles (eight with a low level of colour contrast and eight with a high level of contrast) was mixed randomly and did not change during the experiment. The order of the colour contrast between the motorcycles and the road surface (high level-low level) for the 16 motorcycles was reversed for half of the participants.

The visibility distance of the motorcycle was calculated using the distance (in metres) on the road between the participant and the motorcycle when the participant flashed his/her headlights (note that only responses given while a motorcycle was displayed on the front screens or in the mirrors were taken into account). The greater the distance, the more visible the motorcycle was to the driver (or the earlier the motorcycle could be detected by the driver).

Several ocular indices calculated from the data recorded with the eye tracker throughout the experiment were computed: saccades, eye fixations (on the environment, including elements in the virtual environment and in the participant's car), glances in the mirrors and at vehicles displayed on the screens in front of the participant. Note that these analyses were carried out on the data recorded for 39 participants (20 motorcyclist-motorists and 19 non-motorcyclist-motorists); the data collected for three participants were not of an acceptable quality.

The useful visual field indices were calculated for only 39 participants (18 motorcyclist-motorists and 21 non-motorcyclist-motorists); three participants were unable to take this test for technical reasons.

Results

Analysis of the variance of the visibility distance took into account one between-subjects factor, namely the group of car drivers (motorcyclist-motorists versus non-motorcyclist-motorists), one repeated factor (colour contrast between the motorcycle and the road surface) with two levels (low versus high) and one repeated factor for the arrival position of the motorcycle (a motorcycle which appeared in front of the participant versus a motorcycle which appeared behind the participant).

The mean values of each ocular behaviour index (number and amplitude of saccades, mean duration of eye fixations, number and duration of glances in the mirrors and number and duration of glances at vehicles) and the useful visual field indices (percentage and reaction time) were calculated for the two groups (non-motorcyclist-motorists and motorcyclist-motorists).

The assumptions underlying the ANOVA (that is, normality and uniformity of variance of residuals) were checked using the Shapiro-Wilk and the Levene tests (Zar, 1984). As the ocular behaviour index and reaction times for detecting peripheral signals in the useful visual field test were not normally distributed, the non-parametric Mann-Withney U test was used to analyse inter-group differences. In this chapter, we have presented only significant effects of factors and significant interactions. Comparison of means were carried out using a *post-hoc* LSD Fisher test and means were considered significantly different if the probability of a Type 1 error was less than or equal to 0.05.

Visibility Distance of the Motorcycle

The effect of the arrival position of the motorcycle was significant on visibility distance ($F(1,40)=285$; $p<0.0001$). Drivers detected the motorcycle at a much greater distance when the motorcycle appeared in front of them (average=83.44m, SD=21.1) than when it appeared behind them (average=33.50m, SD=21.9).

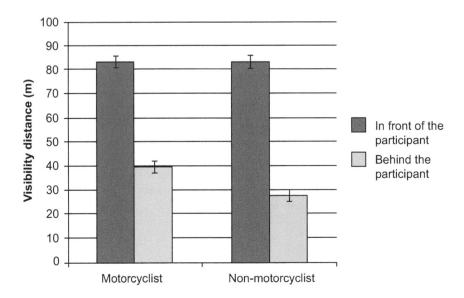

Figure 8.1   Visibility distance (m) as a function of the group of car drivers (motorcyclist-motorists versus non-motorcyclist-motorists) and the arrival position of the motorcycle (behind the participant versus in front of the participant)

The interaction of the arrival position of the motorcycle and the group of car drivers on the visibility distance was significant ($F(1,40)=4.51$; $p=0.04$). Visibility distances are shown in Figure 8.1 above.

A comparison of the averages shows that the group of car drivers affected the visibility distance only when the motorcycle appeared behind the participant. The visibility distance was greater for motorcyclist-motorists than for non-motorcyclist-motorists. These results are consistent with the cognitive conspicuity hypothesis. Car drivers who also had a motorcycle licence and who rode regularly detected the motorcycle when it was further away only when the motorcyclist arrived from behind them.

The interaction of the colour contrast between the motorcycle and the road surface and the arrival position of the motorcycle on the visibility distance was significant ($F(1,40)=5.29$; $p=0.03$). Visibility distances are shown in Figure 8.2.

When the motorcycle appeared in front of participants, they detected motorcycles with a high level of colour contrast at a greater distance than motorcycles with a low colour contrast. When the motorcycle appeared behind participants, the colour contrast had no significant effect. These results are consistent with the sensory conspicuity hypothesis. A significant contrast between the colour of the road surface and the colour of the motorcycle increased the visibility of the motorcycle only when the latter appeared on the front screen.

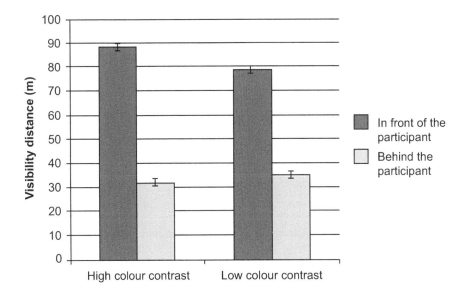

Figure 8.2 **Visibility distance (m) as a function of the colour contrast between the motorcycle and the road surface (high versus low) and the arrival position of the motorcycle (behind the participant versus in front of the participant)**

Useful Visual Field Indices

The percentage of peripheral signals detected did not differ between the two groups. In the current study, motorcyclist-motorists do not appear to have developed a more extensive useful visual field.

Comparison of the two groups' reaction times showed that non-motorcyclist-motorists had longer reaction times than the motorcyclist-motorists. This result supports the useful visual field hypothesis since motorcyclist-motorists are faster at processing the information available in their useful visual field, even if they do not have an extended field. Reaction times are shown in Figure 8.3.

Ocular Behaviour Indices

The effect of driver group was significant on the number of saccades $(F(1,37)=6.03; p=0.02)$. The motorcyclist-motorists carried out more saccades (Figure 8.4).

The difference in saccade amplitude between the two groups was not significant in either the amplitude of the saccades on the vertical plane or in their amplitude on the horizontal plane.

The effect of driver group was significant on the duration of eye fixation $(F(1,37)=7.04; p=0.01)$. Motorcyclist-motorists had a shorter eye fixation duration

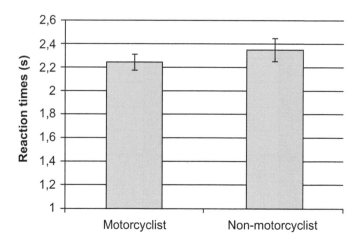

Figure 8.3   Reaction times (s) in the useful visual test for motorcyclist car drivers and non-motorcyclist car drivers

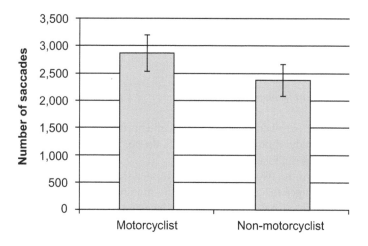

Figure 8.4   Number of saccades for motorcyclist car drivers and non-motorcyclist car drivers

than non-motorcyclist-motorists (Figure 8.5). There was no difference in the number of glances in the mirrors.

The effect of driver group was significant on the duration of glances in the mirrors ($F(1,37)=5.47$; $p=0.02$). Motorcyclist-motorists explored their mirrors more quickly than non-motorcyclist-motorists (Figure 8.6).

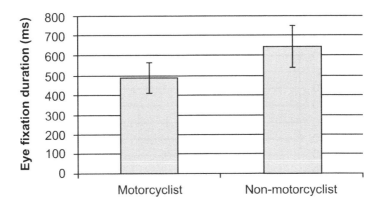

Figure 8.5   Eye fixation duration (ms) for motorcyclist car drivers and non-motorcyclist car drivers

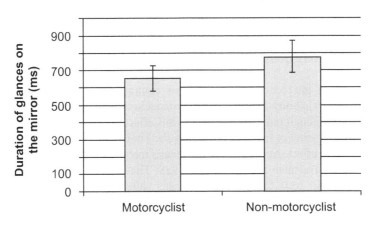

Figure 8.6   Duration of glances in the mirrors (ms) for motorcyclist car drivers and non-motorcyclist car drivers

The effect of driver group was significant on the duration of glances at other vehicles ($F(1,37)=4.25$; $p=0.05$). Motorcyclist-motorists looked at other vehicles more quickly than non-motorcyclist-motorists (Figure 8.7).

These results are consistent with the ocular behaviour hypothesis, according to which car drivers who also have a motorcycle licence and who ride regularly may use specific visual strategies, if we consider that these strategies include a high number of saccades and a rapid capture of information (in the rearview mirrors and at vehicles seen on the screens).

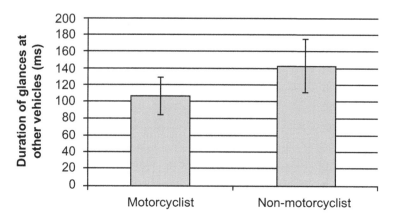

Figure 8.7 **Duration of glances at other vehicles (in ms) for motorcyclist car drivers and non-motorcyclist car drivers**

Discussion

Motorcycle Sensory Conspicuity

The significant interaction between colour contrast and arrival position of a motorcycle on visibility distance led us to conclude that colour contrast has an effect for oncoming motorcyclists and that this effect does not depend on whether drivers have experience in riding a motorcycle. These results therefore illustrate a true bottom-up effect. According to Treisman's integration theory (Treisman and Gelade, 1980; Treisman and Gormican, 1988; Theeuwes, 1991a; Itti and Koch, 2001; Theeuwes, 2004, 2010) and the 'master saliency map' theory (Theeuwes, 1991a, 2004; Itti and Koch, 2001), car drivers develop a saliency map at a pre-attentive level. This takes into account, *inter alia*, the colour and shape (elementary dimensions) of the elements present in the road scene. When the contrast in colour between the motorcycle and the road background is high, car drivers' attention is attracted and directed automatically to locating the salient elements (which may in fact be parts of the motorcycle and/or the motorcyclist) in the road environment by means of a bottom-up mechanism. Car drivers then detect the motorcycle from further away. When the colour contrast between the motorcycle and the road background is low, their attention is thought to be attracted to salient items which are not elements of the motorcycle. The attention search continues with attention passing to the next most salient item (by inhibition of the return mechanism) in order to compare the new item with the target sought. It is believed that a new saliency map is created in case the road image changes (because either the participant or the other vehicles have moved) or if the car driver glances at another location. It may therefore take more time to search for a motorcycle with low colour contrast, and the motorcycle is therefore detected at a shorter visibility distance.

The effect of colour contrast in the current study is in keeping with the results obtained by Hole et al. (1996). Their study concluded that a high colour contrast between the road scene and the motorcyclist's clothing affects visibility in static scenes. The effect of colour contrast observed in our current study also concurs with the notion of an improvement of motorcycle visibility with a high colour contrast when contrast between traffic and a motorcycle is high and the motorcycle is travelling between two lanes of light vehicles (Rogé et al., 2010). We were able to conclude that such an improvement did occur because colour contrast had a significant effect on performance in the motorcycle detection task even when the possibility of anticipating was limited.

Motorcycle Cognitive Conspicuity

Some researchers have suggested that failed and/or late detection of a motorcycle on the road is not exclusively related to its sensory conspicuity (Olson, 1989; Wulf et al., 1989; Hancock et al., 1990; Langham et al., 2002; Magazzù et al., 2006; Comelli et al., 2008). A driver's decisions in relation to a motorcycle or ability to detect a motorcycle could also depend on cognitive conspicuity (Hancock et al., 1990).

One result of this study was that motorcyclist-motorists were found to detect oncoming motorcycles at a greater distance than non-motorcyclist-motorists when motorcyclists came from behind participants. Indeed, the visibility distance was greater for motorcyclist-motorists than for non-motorcyclist-motorists. Visual attention models based on the notion of a 'master saliency map' (which topographically encodes the salience of items in their visual environment) do not enable us to interpret the current results. According to some researchers, saliency calculations for drawing up the saliency map are thought to be more or less restricted to the attentional window of the observers (Theeuwes, 2004; Van der Stigchel et al., 2009; Theeuwes, 2010). A top-down mechanism leads motorcyclist-motorists to move their attentional window in order to examine particular parts of the road scene sequentially because, being motorcyclists themselves, they know that motorcycles ride on specific parts of the road (due to prototypical representations of the scene). This mechanism leads to a low probability that irrelevant salient elements will capture their attention, and the focusing of attention on specific parts of the environment may explain why motorcyclist-motorists detect motorcycles approaching from the rear at a greater distance than do non-motorcyclist-motorists. This interpretation leads to the conclusion that the ability to detect a motorcyclist is closely linked to elements which are specific to drivers, such as their knowledge of the driving behaviour of motorcyclists.

Useful Visual Field and Ocular Behaviour of Car Drivers

The results obtained in the current study do not enable us to conclude that the size of the driver's useful visual field differs between motorcyclist-motorists and non-motorcyclist-motorists. However, the time taken to process the information available in the useful visual field is significantly different in the two groups studied, with motorcyclist-motorists processing it more quickly. They certainly make use of this aptitude to process information rapidly in a task which requires them to divide their attention when driving a car. This result is consistent with the observations made in the analysis of oculometric data recorded during the motorcyclist detection task. The duration of eye fixation was shorter in motorcyclist-motorists, and glances in the rearview mirrors and at other vehicles were also shorter than for non-motorcyclist-motorists. Because motorcyclist-motorists process the information available in their useful visual field more quickly, they required less time to collect information presented in the far periphery in order to identify motorcycles which first appeared in their car's rearview mirror. As a result, motorcyclists arriving from behind were perceived at a greater visibility distance by motorcyclist-motorists.

The fact that motorcyclist-motorists detect information in their useful visual field more quickly is also consistent with the conclusions of earlier research. Horswill, and recently Rosenbloom, assumed that motorcyclists were able to develop an ability to detect, assess and cope with potential hazards (Horswill and Helman, 2003; Rosenbloom et al., 2011). In their study, participants had to imagine that they were driving their usual car, and had to detect hazards in digital video footage of traffic situations projected onto a screen. Comparison of the performances of two groups of car drivers (motorcyclist-motorists and non-motorcyclist-motorists) led the researchers to conclude that motorcyclist-motorists detected hazards significantly more quickly than non-motorcyclist-motorists. Underwood and Chapman also found that motorcyclists had shorter reaction times to hazards and that they classified an object as being a hazard more quickly (Underwood and Chapman, 1998). Finally, Hosking et al. observed a significant decrease in hazard response times as a function of experience in three groups of drivers (inexperienced riders/inexperienced drivers; inexperienced riders/ experienced drivers; experienced riders/experienced drivers) when the hazard perception test was carried out on a motorcycle simulator (Hosking et al., 2010). Moreover, the results revealed that experienced drivers with both a motorcycle and a car licence exhibited a more flexible visual search pattern than the inexperienced riders. Motorcyclists may develop some specific abilities to enable them to detect hazards when they are riding in order to 'survive' on the road, and they may take advantage of these abilities when they drive cars. These abilities, which include being able to process the information in their useful visual field more quickly, may also help them to detect and avoid collisions with motorcycles when they drive a car. An analysis of accident data supports this, since a high proportion of car drivers involved in collisions with motorcyclists do not have a licence to ride a motorcycle, and car drivers who hold a motorcycle licence have been shown to be responsible for fewer motorcycle-car collisions than drivers who do not have one (Hurt et al., 1981; Magazzù et al., 2006; Comelli et al., 2008).

Our current study allowed us to conclude that when the size of the visual field of non-motorcycle-motorists and motorcycle-motorists is identical, the latter group is able to process the available visual information more rapidly.

Conclusion

The results of this study suggest that research on ways of improving the visibility of motorcycles for car drivers should not be reduced to mere modifications of the physical characteristics of motorized two-wheeled vehicles (such as contrast of colour of their helmet, clothes and the body of the motorcycle). This approach would in fact be ineffective in certain cases, particularly when the two-wheeled vehicles appear from behind car drivers and are first seen in their rearview mirrors. In addition, the ability to detect a motorcycle also varies according to a number of characteristics specific to car drivers (such as their riding knowledge and perceptual-cognitive capacities, which are evaluated via their speed in detecting and processing information available in their useful visual field) irrespective of the motorcycle's characteristics (high or low colour contrast with the background).

In future research, it would be interesting to study the effect of a training programme aimed at developing two aspects of visual search related to the detection of motorcycles by car drivers. The first aspect would target the improvement in knowledge of motorcyclists' behaviour, irrespective of whether these behaviours are permitted under the French Highway Code. Riding behaviours could be selected from observations carried out during detailed studies of accidents in which a collision between a car and a motorcycle occurred. Participants would first learn to scan certain specific parts of the road environment according to the characteristics of the situation. The second part of the training programme would include exercises designed to improve the speed at which the information available in the peripheral visual field is processed, using a variety of tasks of increasing difficulty, requiring the division of driver's attention between tasks. The objective of this training programme would be to enhance drivers' ability to detect motorcycles and to anticipate their trajectory correctly (in terms of speed and time to collision) when they appear in their useful visual field.

Acknowledgements

This research was supported by the European Community (Two-Wheeler Behaviour and Safety, project no. 218703). The authors wish to thank the participants, Evgueni Douissembekov, Jacky Robouant, Max Duraz, Daniel Ndiaye and Jean-Paul Rousset (MACIF Rhône-Alpes) for their assistance.

The authors wish to thank Human Factors journal for its permission to reuse parts of the original publication (Rogé et al., 2012) as the source of the information.

References

Ball, K., Beard, B., Roenker, D., Miller, R. and Griggs, D., 1988. Age and visual search: Expanding the useful field of view. *Journal of the Optical Society of America*, 5 (12), 2210–19.

Brown, I.D., 2002. A review of the 'look but failed to see' accident causation factor. London: Department of Transport, Local Government and the Regions.

Cole, B.L. and Hughes, P.K., 1988. Drivers don't search: They just notice. Visual search. D. Brogan. University of Durham, England, 407–17.

Comelli, M., Morandi, A., Magazzù, D., Bottazzi, M. and Marinoni, A., 2008. Brightly coloured motorcycles and brightly coloured motorcycle helmets reduce the odds of a specific category of road accidents: A case-control study. *Biomedical Statistics and Clinical Epidemiology*, 2 (1), 71–8.

Engel, F., 1971. Visual conspicuity, directed attention and retinal locus. *Vision Research*, 11, 563–76.

Engel, F., 1974. Visual conspicuity and selective background interference in eccentric vision. *Vision Research*, 14, 459–71.

Filou, C., Lagache, M. and Chapelon, J. 2005. Les motocyclettes et la sécurité routière en France en 2003, étude sectorielle. Paris, La documentation française.

Hancock, P.A., Wulf, G., Thom, D. and Fassnacht, P., 1990. Driver workload during differing driving maneuvers. *Accident Analysis and Prevention*, 22 (3), 281–90.

Hole, G. and Tyrrell, L., 1995. The influence of perceptual 'set' on the detection of motorcyclists using daytime headlights. *Ergonomics*, 38 (7), 1326–41.

Hole, G., Tyrrell, L. and Langham, M., 1996. Some factors affecting motorcyclists' conspicuity. *Ergonomics*, 39 (7), 946–65.

Horswill, M.S. and Helman, S., 2003. A behavioral comparison between motorcyclists and a matched group of non-motorcycling car drivers: Factors influencing accident risk. *Accident Analysis and Prevention*, 35 (4), 589–97.

Horswill, M.S., Helman, S., Ardiles, P. and Wann, J.P., 2005. Motorcycle accident risk could be inflated by a time to arrival illusion. *Optometry and Vision Science*, 82 (8), 740–746.

Hosking, S.G., Liu, C.C. and Bayly, M., 2010. The visual search patterns and hazard responses of experienced and inexperienced motorcycle riders. *Accident Analysis and Prevention*, 42(1), 196–202.

Hurt, H.H., Ouellet, J.V. and Thom, D.R., 1981. Motorcycle accident cause factors and identification of countermeasures. Los Angeles, California, Traffic Safety Center, University of Southern California, 1.

Ikeda, M., Uchikawa, K. and Saida, S., 1979. Static and dynamic functional visual fields. *Optica Acta*, 26(8), 1103–13.

Itti, L. and Koch, C., 2001. Computational modelling of visual attention. *Nature Reviews Neuroscience*, 2, 194–203.

Langham, M., Hole, G., Edwards, J. and O'Neil, C., 2002. An analysis of 'looked but failed to see' accidents involving parked police vehicles. *Ergonomics*, 45 (3), 167–85.

Leitner, T., Brandstätter, C., Höglinger, S., Bos, N., Berends, E., Yannis, G., Evgenikos, P., Chaziris, A., Broughton, J., Lawton, B. and Walter, M., 2008. European annual statistical report 2008, KfV, NTUA, SWOV and TRL, 64.

Mackworth, N.H., 1965. Visual noise causes vision tunnel. *Psychonomic Science*, 3, 67–8.

Magazzù, D., Comelli, M. and Marinoni, A., 2006. Are car drivers holding a motorcycle licence less responsible for motorcycle – Car crash occurrence?: A non-parametric approach. *Accident Analysis and Prevention*, 38(2), 365–70.

Nagayama, Y., Morita, T., Miura, T., Watanabem, J. and Murakami, N., 1980. Motorcyclists' visual scanning pattern in comparison with automobile drivers. S. o. A. Engineers. Warrendale, PA.

Olson, P.L., 1989. Motorcycle conspicuity revisited. *Human Factors*, 31(2), 141–6.

Rogé, J., Douissembekov, E. and Vienne, F., 2012. Low conspicuity of motorcycles for car drivers: dominant role of bottom-up control of visual attention or deficit of top-down control? *Human Factors* 54(1), 14–25.

Rogé, J., Ferretti, J. and Devreux, G., 2010. Sensory conspicuity of powered two-wheelers during filtering manoeuvres, according the age of the car driver. *Le Travail Humain*, 73(1), 7–30.

Rogé, J., Otmani, S., Pébayle, T. and Muzet, A., 2008. The impact of age on useful visual field deterioration and risk evaluation in a simulated driving task. *European Review of Applied Psychology*, 58, 5–12.

Rogé, J. and Pébayle, T., 2009. Deterioration of the useful visual field with ageing during simulated driving in traffic and its possible consequences for road safety. *Safety Science*, 47, 1271–6.

Rogé, J., Pébayle, T., Campagne, A. and Muzet, A., 2005. Useful visual field reduction as a function of age and risk of accident in simulated car driving. *Investigative Ophthalmology and Visual Science* 46(5), 1774–9.

Rogé, J., Pébayle, T., Hannachi, S.E. and Muzet, A., 2003. Effect of sleep deprivation and driving duration on the useful visual field in younger and older subjects during simulator driving. *Vision Research*, 43(13), 1465–72.

Rogé, J., Pébayle, T., Kiehn, L. and Muzet, A., 2002. Alteration of the useful visual field as a function of state of vigilance in simulated car driving. *Transportation Research, Part F: Traffic Psychology and Behaviour*, 5, 189–200.

Rogé, J., Pébayle, T., Lambilliotte, E., Spitzenstetter, F., Giselbrecht, D. and Muzet, A., 2004. Influence of age, speed and duration of monotonous driving task in traffic on the driver's useful visual field. *Vision Research*, 44(23), 2737–44.

Rosenbloom, T., Perlman, A. and Pereg, A., 2011. Hazard perception of motorcyclists and car drivers. *Accident Analysis and Prevention*, 43(3), 601–4.

Rumar, K., 1990. The basic driver error: Late detection. *Ergonomics*, 33 (10–11), 1281–90.

Sanders, A.F., 1970. Some aspects of the selective process in the functional visual field. *Ergonomics*, 13(1), 101–17.

Scialfa, C., Kline, D., Lyman, B. and Kosnik, W., 1987. Age differences in judgments of vehicle velocity and distance. 31st Annual Meeting of the Human Factors Society, Santa Monica, CA, The Human Factors Society.

Shinar, D., 2007. *Motorcyclists and Riders of Other Powered Two-wheelers: Traffic Safety and Human Behavior*. Amsterdam: Elsevier, 657–94.

Theeuwes, J., 1991a. Cross-dimensional perceptual selectivity. *Perception and Psychophysics*, 50(2), 184–93.

Theeuwes, J., 1991b. *Visual Selection: Exogenous and endogenous control. Vision in Vehicle III*. Amsterdam: North Holland, 53–62.

Theeuwes, J., 2004. Top-down search strategies cannot override attentional capture. *Psychonomic Bulletin and Review*, 11(1), 65–70.

Theeuwes, J., 2010. Top-down and bottom-up control of visual selection. *Acta Psychologica*, 135(2), 77–99.

Tofield, M.I. and Wann, J.P., 2001. Do motorcyclists make better car drivers? Psychological Post-graduate Affairs Group Conference, Glasgow, Scotland.

Treisman, A. and Gelade, G., 1980. A feature-integration theory of attention. *Cognitive Psychology*, 12(1), 97–136.

Treisman, A. and, Gormican, S., 1988. Feature analysis in early vision: Evidence from search asymmetries. *Psychological*, 95(1), 15–48.

Underwood, G. and Chapman, P., 1998. Eye movements and hazard perception ability. Behavioural research in road safety. G. Grayson. Crowthorne: Transport Research Laboratory. VIII, 59–66.

Van der Stigchel, S., Belopolsky, A.V., Peters, J.C., Wijnen, J.G., Meeter, M. and Theeuwes, J., 2009. The limits of top-down control of visual attention. *Acta Psychologica*, 132, 201–12.

Watts, G.R., 1980. The evaluation of conspicuity aids for cyclists and motorcyclists. In Osborne, D.J. and Levis, J.A. (eds) *Human Factors in Transport Research*. Waltham: Academic Press, 203–11.

Wells, S., Mullin, B., Norton, R., Langley, J., Connor, J., Lay-Yee, R. and Jackson, R., 2004. Motorcycle rider conspicuity and crash related injury: Case-control study. *Bristish Medical Journal*, 328(7444), 857–60.

Williams, L., 1982. Cognitive load and the functional field of view. *The Human Factors Society*, 12, 684–92.

Williams, M.J. and Hoffmann, E.R., 1979. Motorcycle conspicuity and traffic accidents. *Accident Analysis and Prevention*, 11(3), 209–24.

Wood, J.M., Dique, T. and Troutbeck, R., 1993. The effect of artificial visual impairment on functional visual fields and driving performance. *Clinical Vision Science*, 8(6), 563–75.

Wulf, G., Hancock, P. and Rahimic, M., 1989. Motorcycle conspicuity: An evaluation and synthesis of influential factors. *Journal of Safety Research*, 20, 153–76.

Zar, J.H. 1984. *Biostatistical Analysis*. Upper Saddle River, NJ: Prentice-Hall.

Chapter 9

Can Drivers' Expectations and Behaviour Around Motorcycles Be Influenced by Exposure?

Vanessa Beanland, Michael G. Lenné and Geoff Underwood

The leading cause of multi-vehicle motorcycle crashes is other road users failing to perceive the motorcycle (ACEM, 2009). This perceptual failure can result from the driver failing to look for the motorcycle (for example, executing a lane change without making a prior head check) or looking but failing to see (see Chapter 3). In both circumstances the motorcycle fails to capture the driver's attention. Several factors influence whether a stimulus will capture attention. Traditionally, researchers differentiated between exogenous and endogenous control of attention. Exogenous attentional capture occurs involuntarily due to 'bottom-up' factors external to the observer, such as stimulus salience, whereas endogenous attentional capture is influenced by internal characteristics of the observer, such as goals or intentions, which are often referred to as 'top-down' influences on attention. Although endogenous attention is a broad concept that encompasses many factors, including one's current goals and recent search history (see Awh et al., 2012), there has been little research examining how it operates in the driving context. The current chapter presents the results of a simulator study designed to explore the extent to which drivers' behaviour around motorcycles can be changed by influencing their endogenous attention, specifically by manipulating their expectations and prior exposure to motorcycles.

Difficulties perceiving motorcycles are commonly attributed motorcycles' low physical salience; for example, the fact that they are significantly smaller than cars and are often dark colours. The vast majority of research regarding motorcycle conspicuity has focused on bottom-up factors that influence drivers' ability to perceive motorcycles. This includes the effectiveness of conspicuity enhancements such as daytime-running lights, novel headlight treatments and reflective or brightly coloured fairings, helmets or clothing (for example, Gershon et al., 2012; Hole et al., 1996; Rößger et al., 2012; Smither and Torrez, 2010; Williams and Hoffman, 1979). Among the various conspicuity enhancements that have been evaluated, daytime-running lights consistently emerge as being the most effective at improving the detectability of motorcycles (for example, Olson et al., 1981; Smither and Torrez, 2010; Thomson, 1980; Williams and

Hoffman, 1979). However, daytime-running lights do not eliminate drivers' perceptual difficulties around motorcycles, they merely reduce them. Moreover, the benefit conferred by daytime-running lights is lessened when other vehicles also use daytime-running lights (Cavallo and Pinto, 2012). This is problematic for motorcyclists given recent legislative changes requiring all new passenger vehicles in the European Union to have compulsory daytime-running lights. Although it is indisputably worthwhile to develop novel treatments that can improve motorcycles' physical salience, there are practical limits on what can be achieved. Given the size differential, it is not physically possible to equate a motorcycle's bottom-up perceptual characteristics with that of a car. For this reason, it is necessary to explore other means of improving drivers' detection of motorcycles.

Research in other domains has suggested that top-down factors such as subjective experience can alter the importance of bottom-up salience. When viewing a photograph or scene observers most commonly fixate on the most physically salient regions first, but if the observers have relevant content expertise (for example, a history student viewing a photograph of artefacts from the US Civil War) then they focus less on physically salient areas and more on semantically meaningful areas (Humphrey and Underwood, 2009). Thus while low salience undoubtedly impairs observers' ability to detect motorcycles, it is possible that the problem of low physical salience could be mitigated by increasing the cognitive salience through altering observers' experience and/or expectations – that is, by manipulating top-down factors that influence endogenous attention.

Within the specific context of drivers' detection of motorcycles, the top-down factor that has arguably received the greatest focus in research literature is the effect of motorcycling experience on drivers' ability to detect motorcycles. 'Dual-drivers' who also ride motorcycles are less likely than non-riders to be at fault in car-motorcycle crashes (Magazzù et al., 2006). It has been suggested that familiarity with motorcycles makes drivers more efficient and cautious when detecting and responding to motorcycles while driving (Crundall et al., 2012; Mitsopoulos-Rubens and Lenné, 2012; Underwood et al., 2011). In eye movement studies dual-drivers tend to spend longer fixating on motorcycles than cars, whereas drivers without a motorcycle licence fixate for similar durations on both cars and motorcycles (Crundall et al., 2012). Fixation duration can be interpreted as a proxy of processing time; thus, the eye movement findings suggest that dual-drivers spend more time processing motorcycles, which is arguably necessary in order to make accurate judgements because motorcycles are smaller and more difficult to perceive than cars. These findings, while limited due to their correlational nature, suggest it may be possible to improve drivers' detection of motorcycles by means other than increasing the vehicles' physical salience, for example by giving drivers greater experience with motorcycles.

There are at least two possible explanations for why dual-drivers are better at detecting motorcycles. The first is that they have an *attentional bias* towards motorcycles, consistent with research in other domains suggesting that individuals with certain characteristics may have a strong attentional bias towards stimuli that are personally meaningful. Past research has indicated, for example, that recreational drug users have an attentional bias towards drug-related items (Jones et al., 2003), insomniacs have a bias towards sleep-related stimuli (Marchetti et al., 2006) and online gaming addicts have a bias towards gaming-related stimuli (Metcalf and Pammer, 2011). Alternatively, they may have an *attentional set* that includes motorcycles, because they have heightened awareness of the hazards facing motorcyclists, whereas for other drivers the attentional set consists exclusively or primarily of four-wheeled vehicles. Attentional set is the notion that observers can 'set' their attention according to specific information that they expect to appear, such as specific objects or colours (Folk et al., 1992). Thus the key difference between attentional bias and attentional set is that observers have an attentional bias towards items with personal relevance, whereas they have an attentional set for items with task-relevance. Although no previous research has explored whether dual-drivers' superior detection of motorcycles is due to attentional bias or attentional set, hazard perception research supports the latter argument. Dual-drivers display more rapid detection of hazards in general, not just motorcycles, suggesting that motorcyclists may adopt a broader attentional set that gives them greater situation awareness while driving (Underwood et al., 2013). From a pragmatic perspective, it is also much easier to modify drivers' attentional set (for example, changing their experiences to give motorcycles greater task-relevance) than it would be to influence their attentional biases. As such, the current study focused on influencing drivers' attentional set by systematically varying their experience with motorcycles, in an attempt to change their subsequent expectations and behaviour around motorcycles.

Extensive research has demonstrated that attentional set forms a powerful top-down influence on what we perceive, by determining our expectations; even a highly salient event, such as woman entering a room and scraping her nails down a chalkboard, can remain unnoticed if it is unexpected (Wayand et al., 2005). Systematic investigation of inattentional blindness, or the failure to detect unexpected objects and events, has revealed that observers are significantly less likely to detect unexpected items that do not match their attentional set (Most et al., 2005; see Chapter 3). Attentional set can also affect drivers' ability to detect hazards during simulated driving. In one study participants were instructed to attend to navigational cues of one colour and ignore another (for example, attend blue, ignore yellow), forcing them to adopt a colour-based attentional set (Most and Astur, 2007). Toward the end of the drive an unexpected motorcycle appeared, which either did or did not match the driver's attentional set colour. Drivers took longer to brake and were significantly more likely to collide with motorcycles that did not match the colour of their attentional set, regardless of the specific colour of the motorcycle.

If we assume that a driver's task is to drive safely, including searching for other vehicles and potential hazards, then drivers theoretically should include motorcycles in their attentional set. After all, motorcycles are a type of vehicle that could appear on the road. In practice, it appears that drivers modify their search strategies depending on what they expect to appear. Most potential hazards can only plausibly appear in a few locations due to the physical limitations of the road infrastructure. This is especially true of large vehicles, which pose the greatest danger in terms of collision and injury risk. From a survival perspective, the most efficient visual search strategy would be for drivers to focus on areas where the most dangerous hazards are most likely to appear and there is evidence that drivers do this at the expense of detecting less common hazards. A Finnish study found that drivers turning right[1] look to their left, where cars could be approaching, but typically fail to look right because cars to their right cannot come into conflict with them (Summala et al., 1996). This can result in crashes with an undetected, less common hazard: cyclists approaching from the right on the footpath next to the road.

The failure to adequately search for less common hazards is consistent with findings from lab-based visual search experiments, in which observers often fail to detect targets that are expected but rarely appear, which has been referred to as the target prevalence effect (Wolfe et al., 2005, 2007). Compared to medium-prevalence targets that appear on 50 per cent of trials, observers are more likely to miss low-prevalence targets that appear on 1–4 per cent of trials, but less likely to miss high-prevalence targets that appear on 96–98 per cent of trials (Schwark et al., 2013; Wolfe et al., 2005, 2007; Wolfe and Van Wert, 2010). Wolfe et al. (2005, 2007) argued that prevalence effects occur because observers adjust their decision criteria depending on target prevalence. Rare targets are unlikely to be present, so observers can make a rapid judgement of 'absent' and be correct 99 per cent of the time. When the targets appear on 50 per cent of trials, there is a 50/50 possibility of being absent so observers require longer to make an accurate judgement. False feedback can alter prevalence effects: telling observers that they have missed a target, even when none was presented, can improve accuracy on subsequent target-present trials and lead to false alarms on target-absent trials (Schwark et al., 2013, 2012). As such it appears that observers adjust their decision criteria based on the perceived prevalence of targets, which is not necessarily the actual target prevalence rate. This in turn suggests that implicit expectations (that is, that the target will not appear, even though the observer has been explicitly instructed to look for it) exert as much influence as explicit expectations.

Effects of target prevalence have been observed in a wide range of search tasks, from very simple arrays of basic geometric shapes (Rich et al., 2008) to highly complex, overlapping images, such as X-ray screens of luggage (Van Wert et al., 2009; Wolfe and Van Wert, 2010). There is also evidence that varying prevalence

---

1   Note that in Finland, vehicles drive on the right side of the road. The right-turn manoeuvre is therefore equivalent to a left-turn manoeuvre in jurisdictions that drive on the left, such as Australia and the UK.

influences detection of targets in real-world tasks, even among observers who have been extensively trained in the search task, such as in cytology screening (Evans et al., 2011). This raises the question of whether drivers' difficulties in detecting motorcyclists could be explained by the fact that motorcycles are an infrequent 'target', constituting around 1 per cent of vehicular traffic in many developed countries including the UK and Australia. In other words, although drivers know that they should search for motorcycles, the fact that motorcycles are rare means that drivers form an implicit expectation that they will not encounter any motorcycles – and consequently they are less likely to detect any motorcycles that do appear.

The current study aimed to investigate whether prevalence effects occur in the context of driving and, if they do occur, how these prevalence effects influence drivers' abilities to detect other vehicles. Participants were required to search for specific target vehicles during a simulator drive. To ensure that observed effects were genuinely due to target prevalence, rather than other stimulus attributes such as the low physical salience of motorcycles, two types of target vehicle were used: motorcycles and buses, which each constitute approximately 1 per cent of traffic in Australia (ABS, 2013). Target prevalence was manipulated both within- and between-subjects, so that half the drivers were exposed to a high prevalence of motorcycles and a low prevalence of buses, while the other half were exposed to a high prevalence of buses and a low prevalence of motorcycles. Prior to completing the main drive, in which observers were explicitly required to detect targets, drivers completed a passive exposure drive in which they were exposed to either a high prevalence of motorcycles (with no buses appearing) or a high prevalence of buses (with no motorcycles appearing). The aim of the exposure drive was to determine whether prevalence effects occur in tasks where observers do not have to actively detect specific targets. It was hypothesised that drivers would be more likely to detect high-prevalence targets and would detect them faster than low-prevalence targets, regardless of vehicle type.

Method

Participants

Forty fully licensed drivers (22 female, 18 male) aged 22–50 years (M=31.9, SD=7.9) participated in exchange for financial compensation. All were experienced drivers (M=13.1 years of licensure, SD=8.0) who drove regularly (M=10.3 hours/week, SD=6.6, range=2–26), typically in urban and suburban areas. All participants provided written informed consent and had normal or corrected-to-normal visual acuity. None had ever held a motorcycle or a bus licence. The ethical aspects of the research were approved by the Monash University Human Research Ethics Committee.

Simulator

Participants drove a medium-fidelity ECA Faros EF-X driving simulator, which consists of a stationary right-hand drive vehicle cab constructed from genuine vehicle parts. The simulator records vehicle parameters including speed, lateral position, braking and acceleration. The visual road environment is projected via three 19-in. LCD screens, which provide 120° horizontal field of view, and realistic traffic sounds are projected through speakers. Two custom buttons were added to the steering wheel to enable participants to respond when they detected targets.

Drive Scenarios

Participants completed two drives: a 7.5 km (4.7 mi.) *exposure drive* followed by a 39 km (24 mi.) *detection drive*. Both drives featured urban roads with one lane in each direction, intersections every 300–500 m and a 60 km/h (37 mph) speed limit. Participants were instructed to obey all road rules, including the posted speed limit and to drive straight across each intersection. Traffic density was kept relatively constant, with other vehicles appearing on the left and right at intersections and oncoming from the opposing direction. The traffic stream consisted entirely of four-wheeled vehicles except for the designated target vehicles, which were motorcycles or buses. The colour (white, grey) and location (left, right, oncoming) of target vehicles were manipulated in a pseudorandom order to ensure that the visual characteristics of the target varied. The exposure drive contained 40 target vehicles. In the 'motorcycle' exposure drive all target vehicles were motorcycles and in the 'bus' exposure drive all targets were buses. The detection drive contained 126 target vehicles. In the 'high motorcycle prevalence' drive 120 targets were motorcycles and 6 targets were buses; in the 'low motorcycle prevalence' drive there were 120 buses and 6 motorcycles. Participants had two buttons located on the steering wheel and were instructed to push the relevant button whenever they detected either a motorcycle or a bus.

Procedure

Participants were told that the purpose of the study was to examine drivers' behaviour and performance in urban traffic environments. For the exposure drive they were instructed to follow the road, drive straight across any intersections and to obey all road rules including traffic signals and posted speed limits. For the subsequent detection drive they were asked to drive normally, but to also use the two buttons on the steering wheel to indicate every time they detected either a motorcycle or a bus. They were informed that reaction time and accuracy were both important, so they should respond as quickly as possible while still obeying the road rules. Before their drives, participants provided demographic information and a brief driving history and completed the 24-item Driver Behaviour Questionnaire (Parker et al., 1995). After their drives, they were asked to indicate their perceived accuracy and difficulty of detecting buses and motorcycles, and to complete open-ended questions indicating: (a) whether anything in particular helped them detect each target type; and (b) whether they used any particular strategies to help them detect targets. Perceived accuracy was recorded as the percentage of targets identified (0–100 per cent). Perceived difficulty was rated on a scale from 1 (very easy) to 5 (very difficult). At the conclusion of the study participants were fully debriefed about the experimental aims.

Data Analysis

There were two 'objective' dependent variables: accuracy (proportion of targets missed) and response time (distance at which the target was detected). There were also two main 'subjective' dependent variables: perceived accuracy and perceived ease of detection. All four dependent variables were analysed using mixed within-between subjects ANOVA with exposure drive (motorcycle, bus) and detection drive (motorcycle high prevalence/bus low prevalence, bus high prevalence/motorcycle low prevalence) as between-subjects factors and vehicle type (motorcycle, bus) as the within-subjects factor. All analyses were evaluated using an alpha level of .05. Partial eta squared (η_p^2) values are reported as measures of effect size, whereby η_p^2=.01 is considered a small effect, η_p^2=.06 is considered a medium effect and η_p^2=.14 is considered a large effect (Cohen, 1988). More detailed analyses for detection distance have been reported elsewhere (Beanland et al., under review), including analysis of how target prevalence interacts with other target features such as salience and location. The following analyses therefore provide only the overall analyses for detection distance, with data for each target type (bus/motorcycle) collapsed across colour and location, and with the addition of the subjective data.

Results

Objective Measures

Misses   Participants were coded as having missed a target if they failed to respond, or responded to it after the vehicle had passed them (this accounted for less than 0.2 per cent of responses). The miss rate was then calculated by dividing the number of targets missed by the total number of targets appearing (6 for low prevalence, 120 for high prevalence), for each vehicle type. The modal and median miss rates were 0 per cent: most observers successfully detected all targets during the drive. Specifically, 68 per cent of observers detected all of the buses and 78 per cent detected all of the motorcycles. Chi-square tests indicated that the probability of missing targets (coded as a dichotomous variable: 0=no missed targets, 1=one or more missed targets) was not significantly influenced by target prevalence rate for either motorcycles, $\chi^2(1)$=1.290, p=.451, or buses, $\chi^2(1)$=1.026, p=.501.

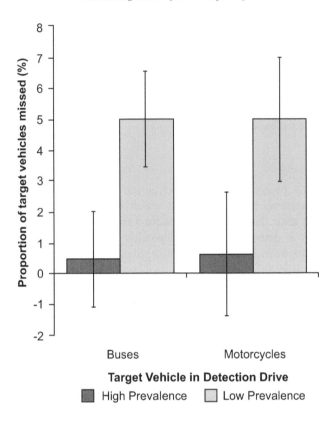

Figure 9.1   Estimated marginal means of miss rates as a function of vehicle type and prevalence

Note: Error bars indicate ±1 standard error.

As shown in Figure 9.1, although there was no difference in the probability of missing targets, miss rates for both vehicle types were higher when they were low prevalence compared to when they were high prevalence. Mixed within-between ANOVA revealed a statistically significant interaction between vehicle type and detection drive, $F(1, 36)=6.689$, $p=.014$, $\eta_p^2=.157$. However, this result should be interpreted cautiously since there were fewer targets in the low-prevalence condition, meaning that missing one target in the low-prevalence condition produces a higher miss rate (1 miss=16.7 per cent miss rate) than missing one target in the low-prevalence condition (1 miss=0.8 per cent miss rate).

Detection distance   There was no significant main effect of exposure drive on detection distance, $F(1, 36)=0.648$, $p=.426$, or detection drive, $F(1, 36)=2.199$, $p=.147$. There was a significant main effect of vehicle type, $F(1, 36)=452.840$, $p < .0005$, $\eta_p^2=.926$: buses were detected farther away than motorcycles. There

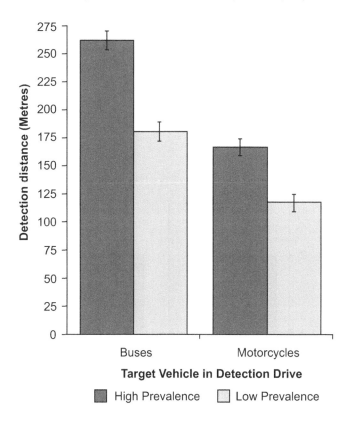

Figure 9.2   Estimated marginal means of detection distance in metres as a function of vehicle type and prevalence

Note: Note that detection distance represents the inverse of response time, so larger values indicate better performance (faster target detection). Error bars indicate ±1 standard error.

was also a significant interaction between vehicle type x detection drive, $F(1, 36)=312.044$, $p < .0005$, $\eta_p^2=.897$, indicating a significant effect of prevalence in the detection drive. As shown in Figure 9.2, both buses and motorcycles were detected significantly farther away when they were high prevalence in the detection drive. Exposure drive did not significantly interact with any variables ($F < 1.2$, $p > .29$, for all comparisons) indicating no significant effects resulting from vehicle prevalence in the exposure drive.

Subjective Measures

Perceived difficulty   Ratings for the perceived difficulty of detecting targets ranged from 1 (very easy) to 4 (difficult) for both buses and motorcycles, with no participants indicating that either target type was 'very difficult' to detect.

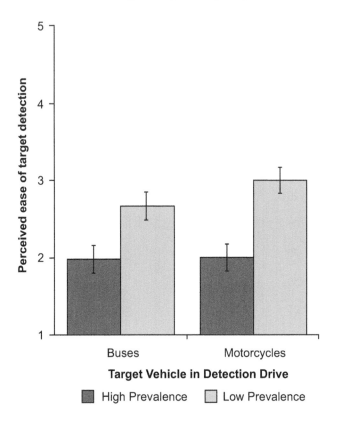

Figure 9.3   Estimated marginal means of perceived difficulty of target detection as a function of vehicle type and prevalence in the detection drive

Note: Note that higher values indicate greater self-reported difficulty in target detection (1=very easy, 5=very difficult). Error bars indicate ±1 standard error.

Mixed within-between ANOVA indicated no statistically significant differences in difficulty ratings as a function of exposure drive, $F(1, 35)=1.301$, $p=.262$, detection drive, $F(1, 35)=0.780$, $p=.383$, or vehicle type, $F(1, 35)=0.993$, $p=.326$. There was a significant interaction between vehicle type and detection drive, $F(1, 35)=22.560$, $p < .0005$, $\eta_p^2=.392$. As shown in Figure 9.3, participants indicated that it was easier to detect high-prevalence vehicles than low-prevalence vehicles, regardless of vehicle type. No other interactions were statistically significant ($F < 1$, $p > .49$, for all comparisons).

Perceived accuracy Data for two participants were excluded from the perceived accuracy analyses because their patterns of answers indicated that they misinterpreted the question (that is, they were reporting the proportion of vehicles

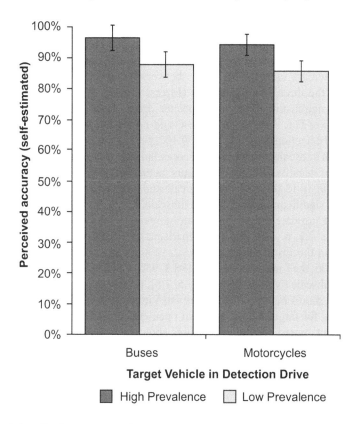

Figure 9.4   Estimated marginal means of perceived accuracy as a function of vehicle type and prevalence in the detection drive

Note: Error bars indicate ±1 standard error.

in the experiment that were buses or motorcycles, rather than the proportion of each target type that they successfully detected). However, the pattern of results remained the same regardless of whether these two participants were excluded. As with perceived difficulty, perceived accuracy showed no statistically significant main effects of exposure drive, $F(1, 34)=0.046$, $p=.832$, detection drive $F(1, 34)=0.000$, $p=.989$, or vehicle type, $F(1,34)=0.448$, $p=.508$. There was a significant interaction between vehicle type and detection drive, $F(1,34)=7.125$, $p=.012$, $\eta_p^2=.173$. As shown in Figure 9.4, participants self-reported that they were more accurate at detecting high-prevalence vehicles compared to low-prevalence vehicles. No other interactions were statistically significant ($F < 1.3$, $p > .26$, for all comparisons). Interestingly, perceived accuracy did not positively correlate with actual accuracy (that is, miss rate). For motorcycles, there was no correlation between perceived accuracy and miss rate, $r=-.147$, $p=.380$, and for buses there was a significant negative correlation, $r=-.602$, $p < .0005$.

Driver behaviour questionnaire (DBQ)   The observed differences in subjective measures cannot be explained by any pre-existing differences in search abilities, driving behaviours or response biases between groups. Mixed within-between ANOVA on the three DBQ subscales (Errors, Lapses, Violations) indicated no statistically significant effects of exposure drive, $F(1, 34)=3.127$, $p=.086$, or detection drive, $F(1, 34)=0.045$, $p=.834$, and no significant interaction between exposure and detection drives, $F(1, 34)=0.102$, $p=.751$. Possible scores ranged from 0–40 on each subscale, where higher scores indicate more frequent self-reported occurrence of negative driving behaviours; actual score ranges were 0–10 for the Error subscale, 0–18 for the Lapse subscale and 0–19 for the Violations subscale. There was a significant main effect of subscale, $F(1.7, 56.1)^2=14.408$, $p < .0005$, $\eta_p^2=.298$, but subscale did not significantly interact with any of the experimental variables ($F < 1.3$, $p > .29$, for all comparisons). Pairwise comparisons indicated that scores on the Error subscale ($M=5.2$, 95% CI [4.4, 6.1]) were significantly lower than both the Lapses subscale ($M=8.8$, 95% CI [7.6, 9.9], $p < .0005$) and the Violations subscale ($M=6.7$, 95% CI [5.5, 7.8], $p=.035$). There was no significant difference in scores between the Lapses and Violations subscales, $p=.055$.

Strategies for target detection – two open-ended questions assessed whether the participants noted any salient features about each target type ('Did anything in particular attract your attention to the [buses/motorcycles]?') and whether they used any specific strategies to facilitate target detection ('Did you use any particular strategies to help you detect the [buses/motorcycles]?') The open-ended responses were coded twice: first, as a dichotomous variable indicating whether they reported any salient features or search strategies; and second, if they did report salient features or search strategies, what these were.

Participants were equally likely to report salient features for buses in the high- and low-prevalence conditions (100 per cent vs. 85 per cent, Fisher's Exact Test, $p=.231$). Salient features reported for buses included: vehicle colour for white buses (63 per cent); larger size (50 per cent); vehicle shape (23 per cent); and windows/windscreens (10 per cent). Participants were also equally likely to report specific search strategies for buses in the high and low-prevalence conditions (75 per cent vs. 60 per cent, $\chi^2(1)=1.026$, $p=.501$). Common search strategies for buses included those based on: characteristic 'blocky' shapes (35 per cent); larger size (33 per cent); specific colours (20 per cent); wheel shape/locations (15 per cent); and vehicle location (for example, at intersections, 10 per cent).

In contrast to the findings for buses, participants were significantly more likely to report salient features for motorcycles when they were high prevalence compared to low prevalence (95 per cent vs. 60 per cent, Fisher's Exact Test, $p=.020$). Salient features reported for motorcycles included: vehicle colour for white motorcycles (40 per cent); distinctive/thinner shape (25 per cent); smaller

---

2   Huynh-Feldt corrections have been applied because Mauchly's test indicated that the assumption of sphericity was violated, $\chi^2(2)=14.945$, $p=.001$, $\varepsilon=.825$.

size (23 per cent); headlights being on (20 per cent); and their position relative to other vehicles (that is, the smaller size of motorcycles meant there was more space between them and other vehicles; 10 per cent). Despite this, participants were equally likely to report specific search strategies for motorcycles in the high and low-prevalence conditions (70 per cent vs. 50 per cent, $\chi^2(1)=1.667$, p=.333). Common search strategies for motorcycles included those based on: smaller vehicles (23 per cent); distinctive/thinner shapes (20 per cent); specific colours (15 per cent); being 'alert' or searching exhaustively (10 per cent); searching for vehicles dissimilar to cars and buses (10 per cent); searching for single headlights (10 per cent); and scanning specific locations (10 per cent).

Discussion

The aim of this study was to explore whether manipulating target prevalence influences drivers' abilities to detect other vehicles. The results support the idea that prevalence significantly affects target detection while driving. Although miss rates were extremely low, particularly in comparison with previous studies of low target prevalence, our results provide some support for the notion that drivers are more likely to miss rare targets in surrounding traffic. The most plausible explanation for why misses were so uncommon is that the nature of the search task gave participants greater opportunity to detect the target. Because drivers were always heading towards targets, they gradually became closer until they passed within a couple of metres of the target vehicle. Assuming that targets were visible from at least 270–450 metres away for motorcycles and buses, respectively, this means that participants had over 15 seconds in which to detect each target. In comparison, visual search tasks using complex real-world stimuli typically find that observers take 2–8 seconds to make a 'target absent' response (for example, Van Wert et al., 2009; Wolfe and Van Wert, 2010). As such, it appears that in our task because participants never had to make a 'target absent' response in order to progress through the experiment, instead of missing targets (that is, making premature 'target absent' responses), the prevalence effect manifested itself as them detected low-prevalence targets significantly later than high-prevalence targets. This explanation is strongly supported by the detection distance data, which reveals that drivers detected high-prevalence vehicles from significantly farther away than low-prevalence vehicles, regardless of vehicle type.

Given that in the real world, both buses and motorcycles are low-prevalence vehicles, our results demonstrate that it is possible to improve detection of vehicles by increasing their prevalence. For buses, increasing prevalence meant that drivers detected them on average 81 metres farther away, or approximately 4.89 seconds earlier at a driving speed of 60 km/h. For motorcycles, high-prevalence vehicles were detected on average 50 metres farther away, or approximately 2.98 seconds earlier at a driving speed of 60 km/h. These findings demonstrate that there can be a dramatic difference between the explicit instructions given to drivers and the implicit expectations that they form based on their own experience of the task. All drivers were explicitly instructed to search carefully for both motorcycles and buses, but our results indicate that their attention was implicitly biased to focus more on the high-prevalence target vehicles. This is consistent with the idea that attentional capture is jointly influenced by multiple factors, specifically physical salience, current goals and selection history (Awh et al., 2012). In an explicit search task, all targets are compatible with current goals (the observer is instructed to search for them), but in the current study they varied in both physical salience (buses are more salient than motorcycles, and were detected from farther away) and selection history (high-prevalence targets have been selected more recently than low-prevalence targets, so they are also detected from farther away).

In addition to the findings from the objective measures of search performance, the subjective self-report measures suggest that drivers are to some extent aware of how prevalence affects their ability to detect targets. In the current study, participants rated high-prevalence targets as significantly easier to detect and believed that they were more successful at detecting high-prevalence targets, compared to low-prevalence targets. Interestingly the correlation between perceived and actual accuracy was negative for buses and non-significant for motorcycles; however, this may be attributable to the fact that there was so little variation in actual accuracy (that is, most participants had a miss rate of 0). For motorcycles, increased prevalence also made drivers more likely to be able to explicitly identify 'salient' features that could facilitate their search for these vehicles; for example, actively searching for shapes that could be the distinctive profile of a motorcycle and rider. Overall this suggests that increasing prevalence was successful in influencing drivers' top-down search mechanisms. As such the current study is one of the first to demonstrate that it is possible to improve drivers' detection of motorcycles by modifying top-down attention factors.

Although these results seem promising, the effects of prevalence were extremely limited in duration. In particular, we only found significant effects of vehicle prevalence during the detection drive; there were no interactions or main effects resulting from previous exposure to a high prevalence of either motorcycles or buses in the initial exposure drive. There are at least two potential explanations for the fact that the exposure drive had no significant effect on subsequent search performance. First, it could be due to the fact that the exposure drive only involved a passive search task, so perhaps participants were not fully attending to the surrounding traffic despite the explicit instructions that they should do so. Anecdotally, a few participants did comment on the high proportion of buses or motorcycles (depending on their condition); for example, one person noted, 'If there were this many buses in real life I wouldn't need to drive because I could take public transport'. However, it remains possible that some drivers failed to register the high prevalence of motorcycles/buses surrounding them, which could explain the null result. A second explanation could be the limited duration of the exposure drive, which took approximately 10–15 minutes to complete. Some previous studies have oscillated target prevalence, where the prevalence rate varied between high and low (for example, Wolfe and Van Wert, 2010). In these circumstances target detection typically improves during and immediately after periods of high prevalence, but decreases again shortly after the prevalence rate drops. As such, it could be that any effects of the exposure drive were eliminated very soon after the start of the detection drive, thus yielding no observable effects.

The current study demonstrates that it is possible to improve drivers' detection of motorcycles (and potentially other hazards, since the effects observed were even larger for buses than motorcycles), which is encouraging, but it is not yet clear how these effects could be maintained long term. Our results suggest that it would not be effective, for example, to implement motorcycle detection 'training' sessions that were one-offs or of brief duration, because the effects would be likely to wear off relatively quickly. However, it should be possible to find a middle ground between the short-term high exposure and short-term effects obtained in the current study, and the long-term effects that appear to arise from personal motorcycling experience (for example, Crundall et al., 2012; Magazzù et al., 2006; Mitsopoulos-Rubens and Lenné, 2012; Underwood et al., 2011, 2012). Longer-term benefits of training, lasting several days to several weeks, have been observed for more general skills such as risk perception (Pradhan et al., 2006) and situation awareness (Stanton et al., 2007; Walker et al., 2009). It should be possible to combine these generic training programs with specific training that focuses on the detection of motorcycles and other vulnerable road users, to equip drivers with search strategies that optimise their visual search performance while driving.

In conclusion, the current study demonstrates that drivers' difficulties in perceiving motorcycles can be attributed to the fact that motorcycles constitute a 'low-prevalence' target on our roads, as well as the fact that they have low physical salience. This in turn provides a new avenue for developing conspicuity treatments, since it is possible to improve detection of low-prevalence targets by manipulating observers' attentional set or their expectations about what will appear. Although previous research on strategies to reduce or eliminate low-prevalence effects has only produced short-term reductions, this does not preclude the possibility of developing future training regimes that can produce longer-term shifts in drivers' attentional sets.

Acknowledgements

This work was supported by an Australian National Health and Medical Research Council Australian-European Union Collaborative Research Grant (ID: 490992). We thank Ash Verdoorn for simulator programming.

References

ABS (Australian Bureau of Statistics). (2013). 9208.0 – Survey of Motor Vehicle Use, Australia, 12 months ended 30 June 2012. Canberra, Australia: Australian Bureau of Statistics.

ACEM (Association des Constructeurs Européens de Motocycles). (2009). In-depth investigations of accidents involving powered two-wheelers (MAIDS): Final report 2.0. Brussels, Belgium: ACEM.

Awh, E., Belopolsky, A.V. and Theeuwes, J. (2012). Top-down versus bottom-up attentional control: A failed theoretical dichotomy. *Trends in Cognitive Sciences*, 16(8), 437–43. doi: 10.1016/j.tics.2012.06.010

Beanland, V., Lenné, M.G. and Underwood, G. (under review). Safety in numbers: Target prevalence affects detection of vehicles during simulated driving. *Attention, Perception, & Psychophysics*.

Cavallo, V. and Pinto, M. (2012). Are car daytime running lights detrimental to motorcycle conspicuity? *Accident Analysis & Prevention*, 49, 78–85. doi: 10.1016/j.aap.2011.09.013

Cohen, J. (1988). *Statistical Power Analysis for the Behavioral Sciences* (2nd ed.). Hillsdale, NJ: Lawrence Erlbaum.

Crundall, D., Crundall, E., Clarke, D. and Shahar, A. (2012). Why do car drivers fail to give way to motorcycles at t-junctions? *Accident Analysis & Prevention*, 44(1), 88–96. doi: 10.1016/j.aap.2010.10.017

Evans, K.K., Evered, A., Tambouret, R.H., Wilbur, D.C. and Wolfe, J.M. (2011). Prevalence of abnormalities influences cytologists' error rates in screening for cervical cancer. *Archives of Pathology and Laboratory Medicine*, 135(12), 1557–60. doi: 10.5858/arpa.2010–0739-OA

Folk, C.L., Remington, R.W. and Johnston, J.C. (1992). Involuntary covert orienting is contingent on attentional control settings. *Journal of Experimental Psychology: Human Perception and Performance*, 18, 1030–1044. doi: 10.1037/0096–1523.18.4.1030

Gershon, P., Ben-Asher, N. and Shinar, D. (2012). Attention and search conspicuity of motorcycles as a function of their visual context. *Accident Analysis & Prevention*, 44, 97–103. doi: 10.1016/j.aap.2010.12.015

Hole, G.J., Tyrrell, L. and Langham, M. (1996). Some factors affecting motorcyclists' conspicuity. *Ergonomics*, 39(7), 946–965. doi: 10.1080/00140139608964516

Humphrey, K. and Underwood, G. (2009). Domain knowledge moderates the influence of visual saliency in scene recognition. *British Journal of Psychology*, 100(2), 377–98. doi: 10.1348/000712608x344780

Jones, B.T., Jones, B.C., Smith, H. and Copley, N. (2003). A flicker paradigm for inducing change blindness reveals alcohol and cannabis information processing biases in social users. *Addiction*, 98(2), 235–244. doi: 10.1046/j.1 360–0443.2003.00270.x

Magazzù, D., Comelli, M. and Marinoni, A. (2006). Are car drivers holding a motorcycle licence less responsible for motorcycle – Car crash occurrence?:

A non-parametric approach. *Accident Analysis & Prevention*, 38(2), 365–70. doi: 10.1016/j.aap.2005.10.007

Marchetti, L.M., Biello, S.M., Broomfield, N.M., Macmahon, K.M.A. and Espie, C.A. (2006). Who is pre-occupied with sleep? A comparison of attention bias in people with psychophysiological insomnia, delayed sleep phase syndrome and good sleepers using the induced change blindness paradigm. *Journal of Sleep Research*, 15(2), 212–21. doi: 10.1111/j.1365-2869.2006.00510.x

Metcalf, O. and Pammer, K. (2011). Attentional bias in excessive massively multiplayer online role-playing gamers using a modified Stroop task. *Computers in Human Behavior*, 27(5), 1942–7. doi: 10.1016/j.chb.2011.05.001

Mitsopoulos-Rubens, E. and Lenné, M.G. (2012). Issues in motorcycle sensory and cognitive conspicuity: The impact of motorcycle low-beam headlights and riding experience on drivers' decisions to turn across the path of a motorcycle. *Accident Analysis & Prevention*, 49, 86–95. doi: 10.1016/j.aap.2012.05.028

Most, S.B. and Astur, R.S. (2007). Feature-based attentional set as a cause of traffic accidents. *Visual Cognition*, 15, 125–32. doi: 10.1080/13506280600959316

Most, S.B., Scholl, B.J., Clifford, E.R. and Simons, D.J. (2005). What you see is what you set: Sustained inattentional blindness and the capture of awareness. *Psychological Review*, 112, 217–42. doi: 10.1037/0033-295X.112.1.217

Olson, P.L., Halstead-Nussloch, R. and Sivak, M. (1981). The effect of improvements in motorcycle/motorcyclist conspicuity on driver behavior. *Human Factors*, 23(2), 237–48. doi: 10.1177/001872088102300211

Parker, D., Reason, J.T., Manstead, A.S.R. and Stradling, S.G. (1995). Driving errors, driving violations and accident involvement. *Ergonomics*, 38(5), 1036–48. doi: 10.1080/00140139508925170

Pradhan, A., Fisher, D. and Pollatsek, A. (2006). Risk perception training for novice drivers: Evaluating duration of effects of training on a driving simulator. *Transportation Research Record: Journal of the Transportation Research Board, 1969*, 58–64. doi: 10.3141/1969-10

Rich, A.N., Kunar, M.A., Van Wert, M.J., Hidalgo-Sotelo, B., Horowitz, T.S. and Wolfe, J.M. (2008). Why do we miss rare targets? Exploring the boundaries of the low prevalence effect. *Journal of Vision*, 8(15). doi: 10.1167/8.15.15

Rößger, L., Hagen, K., Krzywinski, J. and Schlag, B. (2012). Recognisability of different configurations of front lights on motorcycles. *Accident Analysis & Prevention*, 44, 82–7. doi: 10.1016/j.aap.2010.12.004

Schwark, J., Sandry, J., MacDonald, J. and Dolgov, I. (2012). False feedback increases detection of low-prevalence targets in visual search. *Attention, Perception, & Psychophysics*, 74(8), 1583–9. doi: 10.3758/s13414-012-0354-4

Schwark, J.D., MacDonald, J., Sandry, J. and Dolgov, I. (2013). Prevalence-based decisions undermine visual search. *Visual Cognition*, 1–28. doi: 10.1080/13506285.2013.811135

Smither, J.A. and Torrez, L.I. (2010). Motorcycle conspicuity: Effects of age and daytime running lights. *Human Factors*, 52(3), 355–69. doi: 10.1177/0018720810374613

Stanton, N.A., Walker, G.H., Young, M.S., Kazi, T. and Salmon, P.M. (2007). Changing drivers' minds: the evaluation of an advanced driver coaching system. *Ergonomics*, 50(8), 1209–34. doi: 10.1080/00140130701322592

Summala, H., Pasanen, E., Räsänen, M. and Sievänen, J. (1996). Bicycle accident and drivers' visual search at left and right turns. *Accident Analysis & Prevention*, 28, 147–53. doi: 10.1016/0001–4575(95)00041–0

Thomson, G.A. (1980). The role frontal motorcycle conspicuity has in road accidents. *Accident Analysis & Prevention*, 12(3), 165–78. doi: 10.1016/0001–4575(80)90015–9

Underwood, G., Humphrey, K. and Van Loon, E. (2011). Decisions about objects in real-world scenes are influenced by visual saliency before and during their inspection. *Vision Research*, 51(18), 2031–8. doi: 10.1016/j.visres.2011.07.020

Underwood, G., Ngai, A. and Underwood, J. (2013). Driving experience and situation awareness in hazard detection. *Safety Science*, 56, 29–35. doi: 10.1016/j.ssci.2012.05.025

Van Wert, M.J., Horowitz, T.S. and Wolfe, J.M. (2009). Even in correctable search, some types of rare targets are frequently missed. *Attention, Perception, & Psychophysics*, 71(3), 541–53. doi: 10.3758/app.71.3.541

Walker, G.H., Stanton, N.A., Kazi, T.A., Salmon, P.M. and Jenkins, D.P. (2009). Does advanced driver training improve situational awareness? *Applied Ergonomics*, 40(4), 678–87. doi: 10.1016/j.apergo.2008.06.002

Wayand, J.F., Levin, D.T. and Varakin, D.A. (2005). Inattentional blindness for a noxious multimodal stimulus. *American Journal of Psychology*, 118(3), 339–52.

Williams, M.J. and Hoffman, E.R. (1979). Conspicuity of motorcycles. *Human Factors*, 21(5), 619–26.

Wolfe, J.M., Horowitz, T.S. and Kenner, N.M. (2005). Rare items often missed in visual searches. *Nature*, 435(7041), 439–40. doi: 10.1038/435439a

Wolfe, J.M., Horowitz, T.S., Van Wert, M.J., Kenner, N.M., Place, S.S. and Kibbi, N. (2007). Low target prevalence is a stubborn source of errors in visual search tasks. *Journal of Experimental Psychology: General*, 136(4), 623–38.

Wolfe, J.M. and Van Wert, M.J. (2010). Varying target prevalence reveals two dissociable decision criteria in visual search. *Current Biology*, 20(2), 121–4. doi: 10.1016/j.cub.2009.11.066

Chapter 10

Powered Two-Wheelers' Conspicuity: The Effects of Visual Context and Awareness

Pnina Gershon and David Shinar

In the Western world, powered two-wheelers (PTWs) are only a small share of the total motorized traffic. However, they are highly over-involved in the accidents statistics (Shinar, 2007). According to the US National Highway Traffic Safety Administration's 2008 Traffic Safety Annual Assessment (2009), motorcyclist fatalities increased dramatically in the last decade, accounting for 14 per cent of the total US traffic fatalities. Similar statistics were collected in Great Britain, where motorcycles were involved in 14 per cent of all fatal injuries although they accounted for less than 1 per cent of the vehicle population (Clarke et al., 2004). The increase in the number of PTWs (especially the heavy ones), the increase in accidents involving PTWs and the vulnerability of the riders, all together contribute to the concern for motorcyclists' safety (Shinar, 2007). An in-depth study (MAIDS – Motorcycle Accident In-Depth Study) conducted in Europe revealed that 73 per cent of PTWs' accidents occurred at daytime, 90 per cent of them were in clear weather conditions and 85 per cent in light to medium traffic density (ACEM, 2004). Similar findings were obtained in a study conducted in New Zealand, which showed that 64 per cent of the PTWs' crashes occurred in daylight and 72 per cent of these crashes occurred in clear weather (Wells et al., 2004). Given the fact that a large proportion of the accidents involving PTW occurs under fairly safe and favourable environmental conditions, one must ask what are the main factors that contribute to these accidents?

The main contributing factor to the occurrence of accidents involving PTWs is the human factor, which accounted for nearly 88 per cent of all PTW accidents documented in an in-depth, European study (MAIDS) (ACEM, 2009). The results of the MAIDS study indicated that 50 per cent of the total 88 per cent human errors were associated with the other vehicle (OV) drivers (typically drivers of cars or trucks), and in 70 per cent of these cases a perceptual error/failure of the driver was the primary accident cause. Such failures occur most often when crossing a junction without giving the right-of-way to the PTW (Wulf et al., 1989). In 65 per cent of the accidents analysed in an earlier in-depth investigation of motorcycle accidents by Hurt et al. (1981), the drivers of the other vehicles failed to give the PTW the right-of-way. Surprisingly, in many cases, although the PTW was quite close to the junction, the other vehicle's driver failed to detect and distinguish it from the surroundings even when gazing directly at the motorcycle ('look but fail to see' type of failure) (also noted by Pai, 2011; Wulf et al., 1989). Another type of detection error occurs when the other vehicle driver misjudges the speed or distance of the approaching motorcycle (also noted by Cavallo and Pinto, 2011; Pai, 2011). Mannering and Grodsky (1995) offered possible reasons as to why other vehicle drivers do not give the PTW the right-of-way. They claim that car drivers tend to identify other cars as a threat whilst PTWs are not perceived as an endangering, and as such, car drivers do not pay much effort searching for them. Another explanation – specific to intersection accidents – is that drivers who are waiting to enter the junction find it difficult to estimate the time-to-collision with an approaching PTW. Apparently, car drivers overestimate the time that it will take the PTW to enter the junction, probably due to its small perceived dimensions (that is, 'size-arrival illusion') (Horswill et al., 2005).

The term powered two-wheel (PTW) conspicuity is often used to express the likelihood of a PTW being seen by other drivers, and its visual distinction from the surroundings (for example the act of being conspicuous). Though it is not common knowledge that the poor conspicuity of PTWs is one of the main factors contributing to their over-involvement in accidents, especially in fatal ones, there is a significant body of research that documents this (ACEM, 2004). In their review, Wulf et al. (1989) indicated that the PTW size, the surrounding luminance, the PTW's contrast from the background and colour (of both PTW and rider's outfit) affect their detectability. The size and dynamics of the PTW are such that they have lower sensory conspicuity (visual features of an object that make it distinct from its environment), making them more likely to blend in with the surrounding background and more likely to be obscured by cars or blind areas of the other vehicles. In addition, their cognitive conspicuity (distinction of an object based on the observer's experiences and interests) is also poor due to the low exposure frequencies, unexpected locations and unusual behaviours, such as high speeds and 'splitting lanes' – straddling the lane markers (Shinar, 2007; Wulf et al., 1989).

In the past, the sensorial conspicuity of PTWs was improved by the use of daytime-running lights (DRL) that became compulsory for PTWs in many countries (Williams and Hoffman, 1979). DRL can increase the contrast from the background and it is likely that this attention-attracting feature can compensate to some extent for low expectancies and attentional failures (Shinar, 2007). Furthermore, as long as PTWs were the only vehicles using headlights in daytime, DRL provided PTWs with a consistent feature that facilitated their detection and identification by the other vehicle drivers (Brouwer et al., 2004; Olson et al., 1981). However, the relative advantage of the exclusive use of DRL by PTWs is presently diminishing because of the increasing (and in some countries mandatory) use of DRL by passenger cars. It therefore seems necessary to understand and map the intervening factors that influence PTWs' conspicuity. This knowledge can help in finding new ways to enhance PTWs' conspicuity and provide the riders with a unique signature that will make the PTW clearly distinguishable from their background and from other road users.

Another factor that influences a PTW's conspicuity is its distance from the viewer. Hole et al. (2004) showed that the proximity influenced the effectiveness of different aids in increasing PTW conspicuity. The use of headlights, for example, increased the PTW's conspicuity only when it exceeded a certain distance from the viewer. Reflectors, bright clothing and helmets are often used by riders to amplify their conspicuity. Many studies have found that the effectiveness of the clothing and helmet colour, in increasing PTWs' conspicuity, was affected by the contrast with the background (ACEM, 2004; Gershon et al., 2012; Hole et al., 1996; Shinar, 2007). For example, Hole et al. (1996) found that the potency of different conspicuity treatments was primarily affected by environmental factors, and the utility of different conspicuity aids such as headlights and riders' outfits differed according to the environmental characteristics. In a semi-rural environment, headlights improved PTWs' conspicuity whereas in an urban environment their influence was inconsistent. Wells et al. (2004) in a case control study found that wearing a reflective or fluorescent garment decreased the risk of a crash-related injury by 37 per cent. Olson et al. (1981) evaluated the influence of different treatments including headlamp, fluorescent garments and fluorescent fairing on a PTW's conspicuity. They found that both headlamps and fluorescent garments had the potential to increase PTWs' conspicuity, as indicated by bigger gap acceptance. In the MAIDS study, motorcycle riders' use of dark clothing decreased their conspicuity by 13 per cent, though in only 5 per cent of the cases the rider's bright clothing enhanced the PTW's conspicuity (ACEM, 2004).

In this chapter, a series of experiments on PTWs' attention conspicuity and search conspicuity are described. In all of these studies, the PTWs' distance from the viewer, the traffic environments, the times of day, the riders' outfits and helmets and the awareness level of the viewer were all varied in order to assess their influence on PTW attention and search conspicuity. Each type of PTWs' conspicuity (attention conspicuity and search conspicuity) was assessed in two designated experiments: using static pictures and dynamic short video clips of traffic scenes. The aims of the experiments were to identify factors which can increase the ability of a road user to detect a PTW, and to evaluate alternative means of enhancing PTWs' conspicuity in a manner that would be relevant, applicable and efficient in multiple, varied environments.

Attention and Search Conspicuity of a Static PTW

In this section, two experiments on PTWs' attention and search conspicuity are described. In the attention conspicuity experiment, participants were presented with a series of static pictures from different traffic scenarios and were asked to report *what kinds* of motor vehicles were present in each picture. The aim of this experiment was to evaluate the influence of different factors such as PTW distance from the viewer, traffic environments and riders' outfits on the probability that a PTW will be detected by unalerted viewers (that is, when viewers are not primed to expect a PTW within a realistic driving scene). The pictures were staged and captured from real-world driving environments where half of the pictures included a PTW, and half did not.

In a separate experiment, we evaluated search conspicuity. This time the participants were instructed explicitly to search for a PTW in each of the pictures and report its presence or absence, as soon as they reached a decision. The aim of this experiment was to examine search conspicuity of a PTW in terms of the ability and the time needed to detect a PTW by an alerted viewer. Search conspicuity was evaluated in the same traffic environments, the same PTW conspicuity treatments and the same PTW distances from the viewer as in the attention conspicuity experiment.

Attention conspicuity and search conspicuity express two different levels of awareness to the possible presence of a PTW. The different awareness levels can ultimately reflect the effect of 'normal' levels of expectancy (as in the attention conspicuity experiment) and maximal enhancement of expectancy (as in the search conspicuity experiment).

The first two experiments are described in full in Gershon, Ben-Asher and Shinar (2012).

The Effects of Distance, Outfit and Driving Environment on PTW Attention Conspicuity

A total of 66 participants with visual acuity of 6/9 (20/30) or better and a valid driver licence participated in this experiment. The participants were presented with a series of 72 pictures: 36 control pictures without a PTW and 36 with a PTW. Each picture was presented for a fixed duration of 600 milliseconds. Then the participant was asked to report what kinds of motor vehicles were present in the picture. The reporting process was computerized and participants could simply select vehicle types from an on-screen list. The driving scenes that were used in this experiment included:

1. Urban straight road;
2. Urban traffic circle; and
3. Inter-urban road.

Each driving scene included multiple motor vehicles from different categories (for example cars, trucks, motorcycles and so on).

In all of the target pictures (that is, where a PTW was present) the headlights of the PTW were activated, while the use of headlights by the other motor vehicles was not controlled (and mostly off). The influence of PTW the rider's outfit was also evaluated and included: (i) Black clothing with a black helmet; (ii) White clothing with a bright-coloured helmet; and (iii) Reflective vest with a bright-coloured helmet.

The pictures of the driving scenarios used in the experiment were systematically sampled, to obtain pictures of the PTW at different distances from the viewer. The field of view presented in the driving scenes was 28° horizontal, and 19° vertical. The PTW distance from the viewer was represented by the height of the PTW figure measured in pixels (the smaller the height, the farther away the PTW). Four sizes labelled: 'Very Small' (15 pixels), 'Small' (30 pixels), 'Medium' (60 pixels) and 'Large' (120 pixels) were used in the experiment. Given the height of the PTW figures in pixels, and average of 65 cm viewing distance from the screen, the angular size of the 'Very Small' PTW was 0.44°, 'Small' was 0.88°, 'Medium' was 1.76°, and 'Large' was 3.52°.

Each subject first filled out a demographic and driving experience questionnaire. Following that he/she was presented with the instructions and received six training trials (with the ability to perform additional training if he/she needed them). The last phase was the experimental session with 72 trials consisting of 3 (environmental conditions) × 3 (outfits) × 4 (sizes) × 2 (experimental versus control) within subjects design.

The results of this experiment presented in Figure 10.1 shows the detection rates of a PTW as a function of its distance from the viewer, and the PTW rider's outfit. As expected, the detection rates increased as the PTW's distance from the viewer decreased ($\chi2(3)=76.01$, $p<.001$). At the closest distance, the PTW was almost always detected (97 per cent), while at the farthest distance the detection rate was only 26 per cent. Thus, even though the participants were not specifically

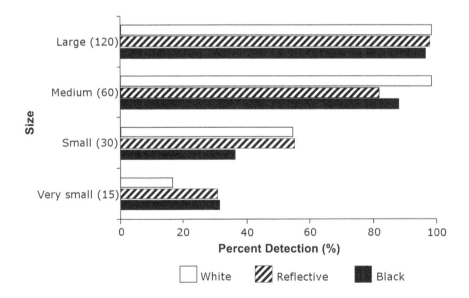

Figure 10.1   Per cent of PTW detection rates as a function of the PTW size and the rider's outfit

instructed to search for a PTW in the pictures, below a certain distance the detection rate was very high and unaffected by the conspicuity treatment (that is, outfit condition).

When combined across all other conditions, the rider's outfit had a marginally significant effect on detection ($\chi2(2)$=5.95, p=.051). However, when the PTW was far from the viewer ('Very Small' and 'Small'), the different outfit conditions and the driving environments affected its detectability ($\chi2(6)$=79.36, p<.001 and $\chi2(6)$=85.85, p<.001, respectively).

Based on the relation between contrast and conspicuity, the following analysis examined the effect of the PTW rider's outfit, and its distance from the viewer on the detection rates in each driving environment separately.

Table 10.1 lists the detection rates of the PTW obtained in each driving environment. When the PTW was distant from the viewer ('Very Small' and 'Small'), the different outfit conditions and the driving environments affected its conspicuity. In the urban roads, where the PTW's background was more multi-coloured (that is, varied) the reflective and white outfits increased its conspicuity compared to the dark clothing condition. In the urban straight road the detection rates of the 'Very Small' and 'Small' PTW rider with reflective outfit were significantly higher compared to the black and white outfits ($\chi^2(6)$=19.27, p=.003 and $\chi2(6)$=32.54, p<.001, respectively). In the urban traffic circle environment the white outfit presented an advantage when trying to detect a PTW. When the PTW was 'Very Small' or 'Small' the detection rates for the white outfit were significantly higher than when the outfit was black ($\chi^2(6)$=13.97, p=.030 and $\chi2(6)$=20.79, p=.002, respectively). A linear contrast analysis yielded a marginally significant difference between the detection rates of the black and reflective outfits in the 'Small' size condition ($\chi^2(6)$=12.28, p=.056).

In contrast to the urban roads, on the interurban road, where the background was only the bright-blue sky, the dark outfit provided an advantage and increased the conspicuity of the PTW. ($\chi2(6)$=82.79, p<.001 and $\chi2(6)$=63.33, p<.001, respectively). In addition, the detection rates were higher for the reflective outfit than the white one ($\chi2(6)$=18.73, p=.004).

These findings are consistent with the results presented in Thomson's review paper (1980) which stated that the higher the contrast between a target and its surroundings the greater the probability of detecting it.

In summary, the results of this experiment reveal the complex relations between the three factors studied: rider's clothing, traffic environment and PTW's distance. As expected, and similarly to previous studies (Hole et al., 1996; Janoff and Cassel, 1971b in Thomson, 1980), the PTW's distance from the viewer had a consistent significant effect on detection regardless of the driving environment and conspicuity treatments. Also similarly to previous findings (for example Hole et al., 1996; Thomson, 1980; Williams and Hoffmann, 1979; Wulf et al., 1989) this study demonstrated how PTW's conspicuity depends on environmental features.

In the following experiment the same set of factors are evaluated when the viewer is alerted and actively searching for the PTW. This can provide a better

Table 10.1   PTW Detection Rates in each of the driving environments

PTWs' Size	PTW Detection Rates (%)			
	Very Small (15)	Small (30)	Medium (60)	Large (120)
Outfit	Urban straight road			
Black	5	6	85	97
Reflective	24*	44*	68	98
White	5	11	97**	100
Outfit	Urban traffic circle			
Black	3	3	79	92
Reflective	14	23	77	95
White	36**	67**	100	95
Outfit	Interurban road			
Black	86***	100	100	100
Reflective	55	98	100	100
White	9	86	98	100

Note: *   The reflective outfit significantly increased detection compared to the black and white outfits.

    **   The white outfit significantly increased detection compared to the black and reflective outfits.

    ***   The black outfit significantly increased detection compared to the white and reflective outfits

understanding of the impact of increased awareness and highlights the difference between search and attention conspicuity.

Effects of Distance, Outfit and Driving Environment on PTW Search Conspicuity

A total of 64 participants with a valid driver licence and visual acuity of 6/9 (20/30) or better participated in this experiment. As in the first experiment (attention conspicuity) the participants were presented with a series of 72 pictures: 36 with a PTW and 36 control pictures without one (the same set of pictures used in the first experiment were used in this experiment). The focus of the experiment was on PTW's search conspicuity, where search conspicuity is defined as the ability to detect a PTW while actively searching for it. Search conspicuity was evaluated in terms of the probability of detecting a PTW and the reaction time to its presence. Note that search conspicuity was evaluated in the same traffic environments, the same PTW conspicuity treatments and the same PTW distances from the viewer as in the attention conspicuity experiment. The participants were instructed to search for a PTW in each of the pictures and report its presence or absence as soon as they reached a decision. They were not asked about any other vehicle types. While viewing the picture, participants were instructed to click on a button as soon as they reached the decision whether a PTW was present in the picture or not. Reaction time (RT) was measured in milliseconds and the maximal presentation time of each picture was 10 sec, after which the picture disappeared. Finally, the experimental platform used to display the pictures in this experiment was identical to the one described in the first experiment.

The detection rates obtained in this experiment reinforced the strong influence that awareness has on the probability of detecting a PTW. The average PTW detection rate of the alerted viewers was 97 per cent across all tested conditions (2195 out of 2268 PTWs were detected). The difference between the detection rates of the largest and the smallest PTW sizes (120 pixels and 15 pixels respectively) was only 8 per cent. Similarly to the results obtained in the first experiment, detection rates of the PTW in the interurban open road environment were the highest and – in parallel – the RT was the shortest. In general, RT increased as the PTW distance from the viewer increased. When the PTW was close ('Large') the average RT required to detect it was approximately 1 sec in all treatments. Therefore, given the importance of early PTW detection, the results of this experiment focused on the ability to detect the PTW when it was distant from the viewer (sizes: 'Very Small' and 'Small'). The average RT to the smallest PTW size was approximately twice as long (1 sec longer) as to the largest size of the PTW. In addition, the average RT to detect a PTW when it was present was 1452 ms (SD= 896.60) across all driving environments, versus 2840 ms (SD=1673.91) to decide that there was no PTW in the picture.

The results presented in Table 10.2 show the influence of the PTW rider's outfit on detection time in each of the driving environments. In general, the environmental characteristics significantly influenced the time needed to detect a PTW. Similarly to the trend established in the first experiment, the reaction times required to detect the 'Very Small', 'Small' and 'Medium' PTWs in the interurban open environment were significantly lower than the ones obtained in both urban roads. A possible explanation is that the urban environment contains a large and diverse repertoire of objects, and each object can act as a distractor and consume some of the mental resources that are required to detect the PTW. The reaction times obtained in the urban environments (straight road and traffic circle) also differed significantly from each other. For the 'Very Small' PTW size the longest RT was in the crowded traffic circle, whereas for the 'Small' to 'Medium' PTW sizes, the straight crowded road yielded the longest RT. The analysis of RT obtained in the straight urban road indicated that the RT to the white and reflective outfits differed in the 'Very Small' and 'Small' sizes. At the farthest distance, the reflective vest significantly decreased the time to detect a PTW, compared to the white outfit condition. However, in the 'Small' size, where the PTW was twice as far, the RT was significantly higher with the reflective outfit than with the white outfit (p=.039). RTs in the urban traffic circle for both 'Very Small' and

Table 10.2   Reaction Time to a PTW in each one of the driving environments and outfits

PTWs' Size	PTW Reaction Time (ms)			
	Very Small (15)	Small (30)	Medium (60)	Large (120)
Outfit	Urban straight road			
Black	2087	1867	1431	1055
Reflective	1888*	2102	1564	1117
White	2393	2063**	1302	1086
Outfit				Urban traffic circle
Black	2573	2328	1119	1250
Reflective	2093***	1788***	1224	1069
White	2253***	1497***	1037	1090
Outfit	Interurban road			
Black	1202 ****	1072	1046	1010
Reflective	1676	1224	969	1052
White	1994	1231	1064	1020

Note: *   The reflective outfit significantly decreased RT compared to the white outfit.

      **   The white outfit significantly decreased RT compared to the reflective outfit.

      ***   The white and reflective outfits significantly decreased RT compared to the black outfit.

      **** The black outfit significantly decreased RT compared to the white and reflective outfits.

'Small' sizes were significantly shorter when the rider was wearing reflective or white outfits, than when wearing a black outfit ($p<.001$ and $p<.001$, respectively). However, this trend is reversed in the interurban road. The RT of a distant PTW wearing black outfit decreased significantly. Linear contrast analysis showed that the RT to the 'Very Small' black outfit was significantly shorter than the RTs of both reflective ($p=.007$) and white ($p<.001$) outfits.

In summary, the second experiment re-emphasizes the strong effects of the environmental characteristics on the time needed to detect a PTW. The reaction times obtained in both cluttered urban environments were significantly longer than the ones obtained in the interurban environment. Similarly to the results obtained in the first experiment and previous studies, the effectiveness of the conspicuity treatment depended on the contrast between the target (PTW) and its background. The reflective and white clothing provided an advantage to the detection of the PTW in the urban environments only when the PTW was far from the viewer. In contrast, in the interurban environment the bright surroundings enhanced the detection of the PTW rider wearing a black outfit. These results are partially consistent with those of Hole et al. (1996) who evaluated the influence of different types of riders' clothing and the use of headlights on PTWs' visibility. They found that in a semi-rural environment with a relatively uncluttered background, dark clothing was superior to bright clothing in terms of PTW visibility. However, their results in the urban environment were not so consistent, and the reaction times to the different clothing conditions were influenced by the presence or absence of the headlights.

Taken together, the results of both the attention and search conspicuity experiments show how the detectability of a PTW is influenced by a blend of visual aspects that are related to the PTW, its rider, the driving environment and the viewers' level of awareness of the possible presence of a PTW. When the PTW was very close and subtended a large visual angle of the observer's visual scene, both unalerted and alerted viewers had near perfect detection rates (97 per cent and 100 per cent respectively). However, at this proximity it may be too late to prevent an accident. Thus, it is important to evaluate potential measures that can increase early detection of the PTW when it is still distant from the viewer. At the farthest distance, the increased awareness (search conspicuity experiment) yielded detection rates three times higher than in the attention conspicuity experiment (92 per cent vs. 26 per cent). Since the level of awareness in the attention conspicuity experiment was not manipulated, it can be regarded as a reflection of PTWs' attention conspicuity at 'normal' levels of expectancy; which are typically low due to the low exposure frequencies. However, the increase in awareness level in the search conspicuity experiment can give an indication of the maximal enhancement possible to detect a PTW in spite of its low exposure frequencies. Hence, increasing awareness to the potential presence of a PTW can counteract the negative influence of its low exposure frequencies.

The use of colour and contrast as means to increase conspicuity is familiar in different fields. The colour attributes of an outfit are frequently used for different and sometimes opposite purposes; such as for camouflage (in army clothing) and to enhance prominence (yellow reflective jackets of road crews). However, the attempt to implement the benefits of contrast and colour on a PTW rider's outfit can be complicated due to the ever changing environments and the PTW's dynamics. As demonstrated in the two experiments, the attention conspicuity and search conspicuity of a PTW rider can be increased by using an appropriate outfit that distinguishes the rider from a given background scenery. Unfortunately, detectability – especially attention conspicuity – is compromised by the perceptual characteristics of the environment that change continuously along a route. Thus, to increase detectability, PTW riders need to be aware of the perceptual aspects of their riding environment. In parallel, the results of the second experiment with alerted viewers demonstrate that other road users (for example car drivers) can improve their detection performance when they increase their level of expectancy and awareness concerning the possible existence of a PTW on the road (as drivers with high expectation obtained nearly 100 per cent detection rates).

In conclusion, the results of the two experiments emphasize the need to direct efforts to enhance the PTW conspicuity in a manner that would be relevant, applicable and efficient in multiple, varied environments. A technique that would increase awareness of all road users to the possible presence of PTW (perhaps as part of the driving learning process, or through advertisements) can dramatically enhance its detectability.

Attention and Search Conspicuity of a Dynamic PTW

In this section, we describe two additional experiments on PTWs' attention and search conspicuity that evolved from and were based on the conclusions and recommendations of the previous experiments. These two experiments are described in full in Gershon et al. (2013).

The following experiments present an attempt to implement the need to provide riders with a unique visual 'signature' that will enhance their attention and search conspicuity. The aim was to create a conspicuity aid that will be less susceptible to environmental variations and enable the other road users to detect the PTW as early as possible. Therefore, we developed a unique Alternating-Blinking Lights System (ABLS) that was based on a movement illusion also known as phi-phenomenon (see Figure 10.2). The specific spacing and timing of the two light sources create the feeling that a movement exists between them, over a range most often described as a moving zone. The ABLS was placed on the rider's helmet (the highest point of the rider) thereby providing the PTW rider with a distinct visual feature (for more details see Gershon et al., 2013).

The third experiment described below focused on attention conspicuity of a dynamic PTW – the ability to detect a moving PTW when the observer's attention is not specifically directed to its possible presence (that is, its ability to draw the observer's attention). The fourth experiment evaluated the search conspicuity of a dynamic PTW – the ability to detect a moving PTW while actively searching for it (Cole and Jenkins, 1984; Gershon, et al., 2012; Wulf, et al., 1989).

These experiments extend the previous experiments in several dimensions. The use of video clips and hence a dynamic and more realistic scene increases the ecological validity of the results. The experiments focused on the complex relations between time of day, driving environment, rider's outfit and helmets and the PTW's distance from the viewer in parallel on both attention and search conspicuity. Thus, in both

Figure 10.2 Frontal and side views of the helmet-mounted ABLS

experiments the same sets of dynamic stimuli were used. The distinction between attention conspicuity and search conspicuity was expressed in two different levels of awareness to the possible presence of a PTW, which can ultimately reflect the effect of 'set' or 'priming' (Tulving and Schacter, 1990).

The novelty of this study lies not only in the evaluation of the ABLS as a conspicuity aid, but rather in the attempt to evaluate the potential of the ABLS to mitigate the differences in detection rate between 'normal' levels of expectancy (reflected in the attention conspicuity experiment) and maximal enhancement of expectancy (reflected in the search conspicuity experiment).

The main hypotheses of both experiments was that the ABLS would provide the rider with a distinct visual feature that will mitigate the influence of the surroundings' background and luminance and increase both PTWs' attention and search conspicuity. We also hypothesized that higher detection rates and shorter reaction times will be obtained at daytime (than at dusk) as the PTW gets closer to the viewer, and when the viewer is cued to search for a PTW (search conspicuity experiment). The following two experiments are described in full in Gershon and Shinar (2013).

Evaluation of PTWs' Attention Conspicuity Using Dynamic Stimulation

This experiment focused on the attention conspicuity of a dynamic PTW. The aim of the experiment was to evaluate the influence of traffic environments, riders' outfits/helmets' features, time-of-day and the distance of the PTW from the viewer on the ability of unalerted viewer to detect a dynamic PTW. The most important aspect of this study is that it also evaluated the potential of the newly developed Alternating-Blinking Light System (ABLS) in increasing PTW attention conspicuity. Twenty students (10 males and 10 females) with visual acuity of 6/9 (20/30) or better and a valid driver licence participated in this experiment. The participants were presented with a series of 96 video clips, 48 with a PTW and 48 control video clips without a PTW. The video clips were randomly presented and included driving environments with multiple motor vehicles from different categories (for example cars, trucks, motorcycles

Figure 10.3   Riders' outfits

and so on). Each video clip was presented for a fixed duration of 800 milliseconds. After the video ended, the participant had to report what kinds of motor vehicles were present in it. The reporting process was computerized and the participants simply had to select the vehicle types from a list on screen. The evaluated driving environments included:

1. Urban road 1, with a speed limit of 50 km/h;
2. Urban road 2 with a speed limit of 50 km/h; and
3. Interurban road with a speed limit of 70 km/h.

Each driving scene was captured at both daytime and dusk. The evaluated PTW riders' outfits included:

1. Black clothing with black helmet;
2. White clothing with a white helmet;
3. Yellow reflective vest with a white helmet; and
4. Yellow, reflective vest with ABLS instrumented white helmet (see Figure 10.3).

Based on the results of previous experiments, PTW attention conspicuity was evaluated when the PTW was at the distances of 140 metres from the viewer and 60 metres from the viewer.

The results summarized in Table 10.3 illustrate the effect of each independent variable on PTW detection. As expected, PTW detection rates increased as its distance from the viewer decreased ($\chi^2(1)=51.8.49$, p<.001). In addition, PTWs detection rates at daytime were significantly higher than at dusk time (approximately 90 per cent vs. 35 per cent) ($\chi^2(1)=162.5$, p<.001). The average detection rates with the ABLS (82 per cent) were the highest among all PTW riders' outfit conditions, and were almost twice as high as the reflective vest condition (48 per cent) ($\chi^2(3)=30.37$, p<.001). Evaluation of the detection rate as a function of the driving environment reveals similar results for all environments, with an average detection rate of 62–63 per cent.

Figure 10.4 shows the detection rates of a PTW as a function of its distance from the viewer and the PTW rider's outfit in both daytime and dusk ($\chi^2(3)=15.24$, p=.002). Analysis of this significant three-way interaction indicated that at dusk the detection rates decreased as the distance of the PTW from the viewer increased, regardless of the type of rider's outfit. However, at daytime the

Table 10.3   Detection rates of a dynamic PTW

Distance from the viewer	Detection rate (%)
140(m)	51
60 (m)	74
Time-of-Day	**Detection rate (%)**
Daytime	89
Dusk	35
PTW outfit	**Detection rate (%)**
Black	62
White	65
Reflective	48
ABLS	82
Driving environment	**Detection rate (%)**
Urban road 1	62
Urban road 2	63
Inter-urban road	63

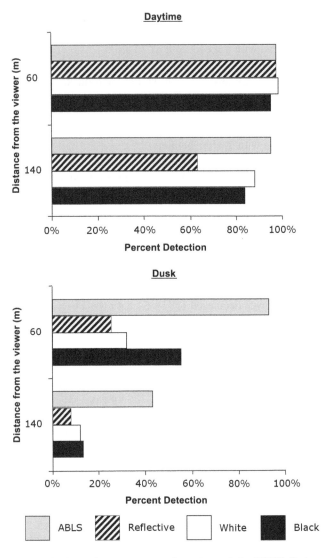

Figure 10.4   PTW detection rates as a function of the PTW distance from the viewer and the rider's helmet/outfit colour, at daytime (top) and at dusk (bottom)

detection rates of both black and ABLS outfits were not affected by the rider's distance from the viewer (p=0.07, p=0.53 respectively). The low levels of luminance at dusk decreased PTW detection rates in all driving environments, regardless of the conspicuity treatments, but the ABLS-mounted helmet was less susceptible to the decrease in the level of luminance from daytime to dusk. The detection rates of the ABLS established in daytime and dusk, when the PTW was close (60m), did not differ significantly from each other (p=0.2).

Based on the relation between the riding environment, contrast and conspicuity, the following analysis examined the effect of the PTW rider's outfit and its distance from the viewer on the detection rates at daytime and dusk in each driving environment separately.

An analysis of the detection rates obtained in the urban road 1 environment indicated that both rider's outfit and distance from the viewer had significant main effects ($\chi^2(3)=33.34$, p<.001 and $\chi^2(1)=15.42$, p<.001, respectively). Moreover, at dusk in both urban environments detection rates were higher with the ABLS. At both distances the detection rates with the ABLS were significantly higher than the ones obtained with the other conspicuity treatments. In the interurban environment, the detection rates were affected by all of the variables: time-of-day, rider's outfit, and the PTW distance from the viewer ($\chi^2(1)=64.06$, p<.001; $\chi^2(3)=12.35$, p=.006; $\chi^2(1)=28.72$, p<.001 respectively), and by the interaction between the time-of-day and PTW distance from the viewer ($\chi^2(1)=5.18$, p=.023). In addition to the overall lower rates of detection at dusk, the effect of the decrease in PTW distance on the detection rates was more dramatic at dusk than at daytime.

In summary, the findings of the current experiment demonstrate the strong influence of the environmental factors (time-of-day and driving environment) on attention conspicuity. Detection rates were first and foremost influenced by the surrounding luminance levels. The complexity increases even more with the inevitable need to take into consideration the influence of the driving environment. The results indicated that the environmental characteristics influenced the likelihood that the viewer will detect the PTW. Similar to previous findings, including the ones presented above (Wulf et al., 1989; Hole et al., 1996; Gershon et al., 2012), these results emphasize the need to develop conspicuity treatments that will be as unsusceptible as possible to changing environmental conditions. As such, the ABLS demonstrated a positive influence on PTWs' attention conspicuity by moderating the masking effects of the environment, mainly at dusk, and in urban environments.

Evaluation of PTWs' Search Conspicuity Using Dynamic Stimulation

In this experiment search conspicuity was manipulated by altering the viewer's level of awareness to the possible presence of a PTW in the driving scenery. A total of 20 participants with a valid driver licence and visual acuity of 6/9 (20/30) or better participated in this experiment. As in the previous experiment (attention conspicuity of dynamic PTW), participants were presented with the same set of 96 video clips; 48 with a PTW and 48 control video clips without a PTW. The instructions were to search for a PTW in each of the video clips and to report its presence or absence by pressing the space bar as soon as the participant reached a decision. This stopped the video and cleared the screen. Reaction time (RT) was measured in milliseconds and the maximal presentation time of each video clip was set at 2 sec, after which the video clip ended.

The results of this experiment are presented in Figure 10.5, which shows the detection rates of the dynamic PTW. The average PTW detection rates of the alerted viewers was 77 per cent across all tested conditions (735 out of 960 PTW presentations were detected). At daytime, the detection rates of the alerted viewer were nearly 100 per cent in all riders' outfit conditions. However, at dusk the detection rates were significantly affected by the rider's outfit ($\chi^2(3)=33.23$, p=.001) and by the interaction between PTWs' distance from the viewer × rider's outfit ($\chi^2(3)=26.97$, p=.001). The detection rates of the rider wearing a black outfit

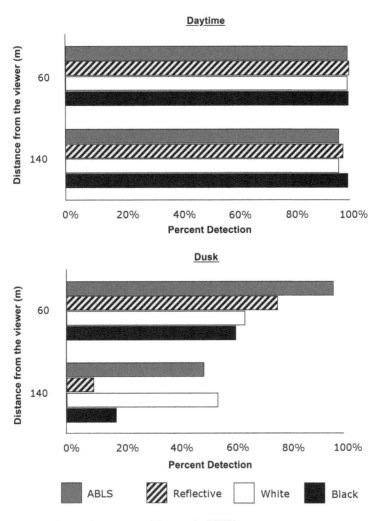

Figure 10.5 Detection rates of dynamic PTW

Table 10.4   Detection rates in the two levels of expectancy at dusk time

	PTWs' Detection Rates at Dusk (%)		
	Attention conspicuity	Search conspicuity	Difference (Δ)
Black	34	40	6
White	22	65	43
Reflective	17	43	27
ABLS	68	73	5

were substantially lower than the ones obtained in the ABLS condition at both distances (140m and 60m, p=.001 and p=.001).

Table 10.4 presents dusk time detection rates at the two levels of expectancy: 'normal' levels of expectancy (based on the attention conspicuity experiment) and maximal enhancement of expectancy (based on the search conspicuity experiment). The analysis of viewers' expectancy levels and the rider's outfit yielded significant main effects of the rider's outfit, expectancy levels and a significant interaction between the two variables ($\chi^2(3)$=90.41, p=.001; $\chi^2(1)$=37.57, p=.001; $\chi^2(3)$=20.55, p=.001 respectively). In the attention conspicuity study the detection rates of riders wearing white and reflective outfits significantly increased as the level of expectancy increased (p=.001 and p=.001 respectively). In contrast, in the search conspicuity condition with the high expectancy level, the detection rates for the black outfit and the ABLS were not significantly different from their detection rates in the attention experiment (p=.35 and p=.39 respectively). It seems that the outfit of the rider moderated the influence of the expectancy level at dusk. In the two extreme conditions of reflectance – when the rider wore a black outfit or when the rider was equipped with the ABLS – increasing drivers' awareness to the possible presence of a PTW did not increase detection rates. However, in the intermediate conditions of reflectance (white and reflective outfits) the increase in level of awareness dramatically increased PTWs' detection rates.

One possible explanation for these results is that the effects of expectancy are limited and have fixed boundaries that are determined by the rider's conspicuity. The black outfit represents the less conspicuous outfit at dusk, and even at a maximum level of awareness – as illustrated in the search conspicuity experiment – cannot overcome the difficulty of detecting a black object in a low luminance environment. Thus, although at dusk the attention conspicuity of the black outfit was higher than that of the white and reflective outfits, the search conspicuity – with the high expectancy – did not increase its detection rates (as opposed to the other two outfits). In contrast, the ABLS increased PTWs' detection rates in both experiments and diminished the difference in the detection rates between the 'normal' and the 'maximal' levels of expectancy (as reflected in the results of the attention and search conspicuity experiments respectively).

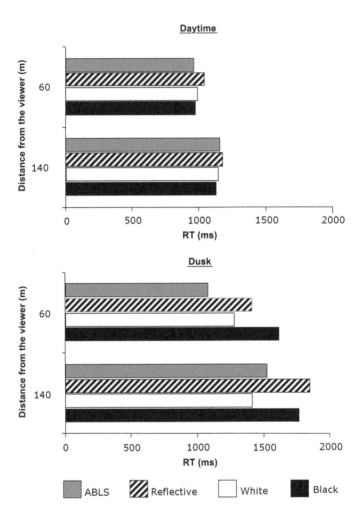

**Figure 10.6   Average RT to detect a PTW as a function of distance
and outfit**

Figure 10.6 shows the average RT as function of the time-of-day, PTW distance
from the driver, and the different riders' outfits. In general, the RTs increased when
the background luminance decreased (that is, at dusk), and the shortest RTs were
obtained in the ABLS condition. The interaction between time-of-day and rider's
outfit showed that the ABLS moderated the increase in RT when the ambient
luminance decreased compared to all other outfits. As expected, the most drastic
increase in RT between daytime and dusk obtained in the black outfit condition.

Figures 10.7 and 10.8 illustrate the average RT obtained in the second urban
environment. A significant three ways interaction was found between time-of-

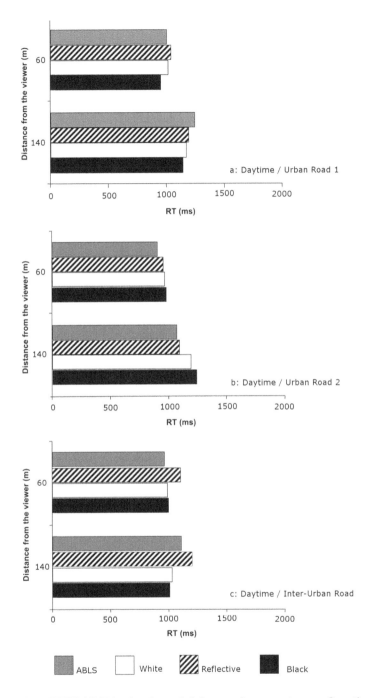

Figure 10.7   PTWs' RT in the three driving environments, as a function of
PTW distance from the viewer and rider's outfit (daytime)

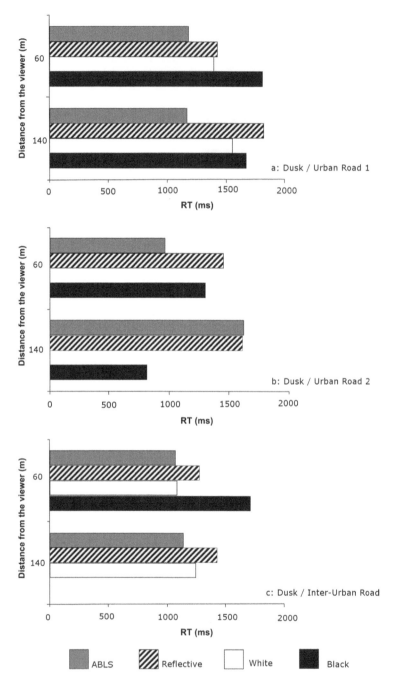

Figure 10.8   PTWs' RT in the three driving environments, as a function of PTW distance from the viewer and rider's outfit (dusk)

day x rider's outfit x PTW's distance from the viewer [F(2, 203)=12.02, p<.001]. At dusk, the white outfit completely masked the PTW, making it essentially undetectable (and therefore having no RT data). However, the RT to the ABLS at dusk was the lowest, especially when the PTW was close to the viewer.

The RTs in the interurban driving environment (Figure 10.7 and 10.8) were significantly affected by the PTW distance from the viewer [F(1, 240)=9.38, p<.026] and by the interaction between time-of-day and rider's outfit [F(3, 240)=11.81, p<.001]. At dusk, the black outfit took the longest to detect, so much so that at the far distance the black-wearing PTW rider was not detected at all, and at the close distance he elicited the longest RT. On the other hand, the RT to ABLS was the shortest in daytime and dusk and at both distances from the viewer.

In summary, the detection rates of the dynamic stimulus revealed that two seconds were more than enough for an alerted viewer to detect a PTW at daytime luminance condition. However, at dusk the detection rates were susceptible to the environmental conditions, and were substantially lower, even when the subjects were actively searching for the PTW. Interestingly, when comparing the detection rates obtained from the search and attention conspicuity experiments we found that the type of outfit the rider was wearing moderated the influence of expectancy level at dusk. This result has an important implication: increasing the viewer's awareness cannot serve as the only conspicuity countermeasure as it is still affected by the visual aspects of the rider, as was illustrated when the rider wore a black outfit or with the ABLS aid. The RT analysis indicated that there was a general increase in RT when the background luminance decreased. However, in congruence with the results obtained in the previous experiment, the ABLS was less susceptible to the reduced ambient illumination, and mitigated the increase in RT at dusk. When the PTW was close to the viewer, the ABLS yielded the lowest RT across all driving environments. Finally, the ABLS was the most potent conspicuity treatment of the ones evaluated in this experiment, and provided the PTW rider with a consistent conspicuity advantage.

The results of the two experiments demonstrate how alerted road users (for example car drivers) can improve their detection performance by increasing their level of expectancy to the potential presence of a PTW on the road. In addition, the experiments with the dynamic views underscore the limits of relying on increased awareness alone. It seems that under less favourable environmental conditions (for example, dusk), increasing the viewer's awareness cannot serve as the only conspicuity countermeasure, as it is still subject to the visual aspects of the rider (as was illustrated when the rider wore a black outfit or with the ABLS aid).

A PTW's rider conspicuity can be increased by using an appropriate outfit or conspicuity aid that distinguishes the rider from a given background scenery. In both attention and search conspicuity experiments, the ABLS yielded the strongest and most consistent positive effect on the detection of a PTW in all driving environments. The advantage of the ABLS was mainly at dusk, when the luminance levels were low. In general, the ABLS was less susceptible to the environmental changes (that is, luminance), increasing PTWs' detection rates and moderating the increase in the reaction time to the presence of a PTW.

General Discussion and Conclusions

This chapter described four complementary experiments that illustrated the complex interactions between the various factors contributing to PTWs' attention and search conspicuity. The experiments presented here were designed and conducted in a hierarchical manner, proceeding from the simplest to the most complex. The first two experiments addressed the issues of attention and search conspicuity using static stimuli and manipulated the riders' outfits and helmet colours. The third and fourth experiments focused on dynamic stimuli in dynamic environments, and introduced a novel means of enhancing the PTW's conspicuity with a phi-phenomenon effect created on the top of the helmet – the alternating blinking light system (ABLS). The aims of all the experiments were to identify factors that can increase the ability of a road user to detect a PTW, and to evaluate means for enhancing PTWs' conspicuity in a manner that would be relevant, applicable, and efficient in multiple, varied environments.

The attempt to exploit the benefits of contrast and color on a PTW rider's outfit was found to be complicated due to the ever-changing environments and the PTW dynamics. As demonstrated in the first two experiments, the attention and search conspicuity of a PTWs' rider can be increased by using an appropriate outfit that distinguishes the rider and the PTW from a given background scenery. Unfortunately, conspicuity is compromised by the perceptual characteristics of the environment that change continuously along a route. Thus, to increase detectability, PTW riders need to be aware of the perceptual aspects of their riding environment, and adapt to it; a practically impossible task. In parallel, the results of the second and fourth experiments with alerted viewers demonstrated that other road users (e.g. car drivers) can improve their detection performance when they increase their level of expectancy and awareness concerning a possible existence of a PTW on the road.

In the third and fourth experiments, a newly developed lighting 'signature' was introduced and evaluated. The alternating blinking lights system (ABLS) – a unique and novel display of two helmet-mounted lights that created an illusion of movement (known as phi-phenomenon) – was fixed on the top of the rider's helmet (the highest point of the PTW's rider). In addition to the assessment of different driving environments we also evaluated the influence of expectancy on the probability of detecting a PTW. Because the level of awareness in the attention conspicuity experiment was low it can be regarded as a reflection of PTWs' attention conspicuity at 'normal' levels of expectancy; which are typically low due to the low exposure frequencies of PTWs – at least in most of the Western world (Cavallo and Pinto, 2011; Gershon et al., 2012; Pai, 2011; Shinar, 2007; Wulf et al., 1989). However, the increase in the level of awareness in the search conspicuity experiment can provide an estimate of the 'maximal' enhancement possible to detect a PTW in spite of the low exposure frequencies. Taken together, the results of the four experiments indicate that alerted road users (for example car drivers) can improve their detection performance by increasing their level of expectancy to the potential presence of a PTW on the road. Differently from the first two experiments described above, the current search conspicuity experiment points out the limits of relying on increased awareness alone. It seems that under less favourable environmental conditions (for example, dusk) increasing the viewer's awareness cannot serve as the only conspicuity countermeasure, as it is still subject to the visual aspects of the rider (as was illustrated when the rider wore a black outfit or with the ABLS aid).

In both attention and search conspicuity experiments, the ABLS had the strongest and most consistent positive effect on the detection of a PTW in all driving environments. The advantage of the ABLS was mainly apparent at dusk, when the luminance levels were low. In general, the ABLS was found to be less susceptible to the environmental changes (that is, luminance), increasing PTWs' detection rates and moderating the increase in the reaction time to the presence of a PTW.

In conclusion, it seems that the best strategy to increase PTWs' conspicuity should take into account both attention and search conspicuity aspects. Therefore, it should involve both the PTW riders and the other vehicle drivers. From the attention conspicuity perspective, the riders need to take into consideration the perceptual aspects of their riding environment, and incorporate conspicuity aids, such as the ABLS, that are less susceptible to the environmental influences and provide the rider with a unique visual signature. In parallel, the expectancy level of other vehicle driver (search conspicuity) should be increased and maintained at a high level of awareness of the possible appearance of a PTW.

Acknowledgments

Work on this study was supported in part by the European Commission 2BeSafe project of the 7th European Framework.

References

ACEM (2004). MAIDS: In-depth investigations of accidents involving powered-two-wheelers. Association de Constructeurs Européens de Motorcycles (ACEM), Brussels, BG. Retrieved from: http://www.maids-study.eu

ACEM (2009). MAIDS: In-depth investigations of accidents involving powered two-wheelers. Multivariate Analysis of MAIDS Fatal Accidents. Association de Constructeurs Européens de Motorcycles (ACEM), Brussels, BG. Retrieved from: http://www.maids-study.eu/pdf/MAIDS_Multivariate_Analysis.pdf

Brouwer, R.F.T., Janssen, W.H. Theeuwes, J. Duistermaat, M. and Alferdinck, J.W.A.M (2004). Do other road users suffer from the presence of cars that have their daytime-running lights on? (Research Report). Soesterberg, the Netherlands: TNO.

Cavallo, V. and Pinto, M. (2011). Evaluation of Motorcycle Conspicuity in a Car DRL Environment. Proceedings of the 6th International Symposium on Human Factors in Driver Assessment, Training, and Vehicle Design, Lake Tahoe, CA, US.

Clarke, D.D., Ward, P., Bartle, C. and Truman, W. (2004). In-depth Study of Motorcycle Accident. Road Safety Research. Report No. 54. Department for Transport, London.

Gershon, P., Ben-Asher, N. and Shinar, D. (2012). Attention and search conspicuity of motorcycles as a function of their visual context. *Accident Analysis and Prevention*, 44, 97–103.

Gershon, P., and Shinar, D. (2013). Increasing motorcycles attention and search conspicuity by using Alternating-Blinking Lights System (ABLS). *Accident Analysis and Prevention*, 50, 801–10.

Harrell, F.E. (2001). *Regression Modeling Strategies with Applications to Linear Models, Logistic Regression and Survival Analysis*. New York: Springer Science & Business Media.

Harrison, W.A. (2005). A demonstration of avoidance learning in turning decisions at intersections. *Transportation Research Part F*, 8, 341–54.

Hole, G.J., Tyrrell, L. and Langham, M. (1996). Some factors affecting motorcyclists' conspicuity. *Ergonomics*, 39, 946–65.

Horswill, M.S., Helman, S., Ardiles, P. and Wann, J.P. (2005). Motorcycle accident risk could be inflated by a time to arrival illusion. *Optometry and Vision Science*, 82(8), 740–746.

Hung, B. and J. Preston (2004). *A Literature Review of Motorcycle Collisions*. Transport Studies Unit, Oxford University, Oxford, UK.

Hurt, H.H. Jr, Ouellet, J.V. and Thom, D.R. (1981).Motorcycle accident cause factors and identification of countermeasures: volume 1: technical report. National Highway Traffic Safety Administration. Report DOT-HS-805–862. US. Department of Transportation, Washington DC.

Jama, H.H., Grzebieta, R.H., Friswell, R. and McIntosh, A.S. (2011). Characteristics of fatal motorcycle crashes into roadside safety barriers in Australia and New Zealand. *Accident Analysis & Prevention*, 43, 652–60.

Mannering, F.L. and Grodsky, L.L. (1995). Statistical Analysis of Motorcyclists' Perceived Accident Risk. *Accident Analysis and Prevention*, 27, 21–31.

NHTSA (2009). Trafic Safety Facts 2008 data-Motorcycles. National Highway Traffic Safety Administration. Report HS811 159. US Department of Transportation Washington DC.

Olson, P.L., Halstead-Nussloch, R. and Sivak, M. (1981). The effect of improvements in motorcycle/motorcyclist conspicuity on driver behavior. *Human Factors*, 23(2), 237–48.

Pai, C.W. (2011). Motorcycle right-of-way accidents – a literature review. Accident *Analysis and Prevention*, 43, 971–82.

Posner, M.I. and Dehaene, S. (1994). Attentional networks. *Trends in Neuroscience*, 17, 75–9.

Shinar, D. (2007). *Traffic Safety and Human Behavior. Chapter 16: Motorcyclists and Riders of Other Powered Two-Wheelers*, 657–94. Elsevier, Oxford, UK.

Thomson, G.A. (1980). The role frontal motorcycle conspicuity has on road accidents. *Accident Analysis and Prevention*, 12, 165–78.

Tulving, E. and Schacter, D.L. (1990) Priming and human memory systems. *Science*, 247(4940), 301–6.Wells, S. Mullin, B. Norton, R. Langley, J. Connor, J. Lay-Yee, R. and Jackson, R. (2004) Motorcycle rider conspicuity and crash related injury case-control study. *British Medical Journal*, 328, 857–63.

Williams, M.J. and Hoffmann, E.R. (1979). Motorcycle Conspicuity and Traffic Accidents. *Accident Analysis and Prevention*, 11, 209–24.

Wulf, G., Hancock, P.A. and Rahimi, M.(1989). Motorcycle conspicuity: an evaluation and synthesis of influential factors. *Journal of Safety Research*, 20, 153–76.

PART IV
Implications Drawn from the Case Studies

Chapter 11

Summarised Assessment of the Results on Motorcycle Conspicuity

Lars Rößger, Michael G. Lenné and Stéphane Espié

Introduction

The present volume focuses on several aspects related to the perception of PTWs by other road users. In order to provide a summarised overview about the main findings from the case studies, it is useful to restructure the outcomes according to the basic behavioural taxonomy introduced earlier, and with respect to the classification of typical errors (see Chapter 3) conducted by other road users in the interacting course with PTW riders. Based on the three key behaviours emphasized by Crundall et al. (2008), two main error categories were differentiated in Chapter 3: *detection errors*, referring to road users' failure to look at and to properly detect a PTW, and *decision errors*, referring to the inadequate appraisal of an oncoming PTW in terms of its speed and/or its path. The results can be further discussed on a second level with respect to the impact of PTWs' visual characteristics as bottom-up component within the perception processes, and on the other hand, with respect to the characteristics of the observer and decision-maker operationalizing the top-down influences, and, respectively, the cognitive factors for the road user interacting with the PTW in traffic.

In addition to a summary of the main results, the present chapter discusses and compares the various methods presented within this volume to assess the role of conspicuity in PTW safety, focusing also on the strengths and drawbacks of each methodology. The studies presented within this volume can be broadly distinguished in terms of their experimental representation of the investigation's focus: several studies used screen experiments to consider research questions related to PTWs' conspicuity, while others implemented simulations of the driving task and/or the traffic situation in order to address research questions. Beyond the scope of the present case studies, further experimental set-ups would be also conceivable, such as the consideration of PTWs' conspicuity in field experiments and real traffic situation. Aside from the ecological validity of the experimental set-up, each type of implementation will also have certain effects on the suitability of methods that can be used to measure conspicuity-related effects and/or on the possibility to experimentally manipulate certain treatments while controlling for influences due to secondary variance.

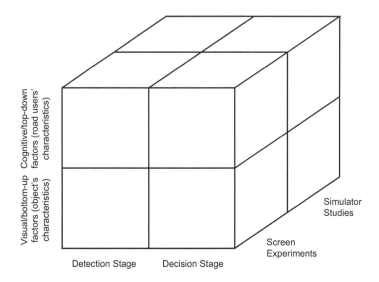

Figure 11.1   Schematic overview of main aspects in classification of case studies' results

Deriving from the points mentioned above, we will reconsider the results according to three axes representing:

a. the focused key behaviour/error stage of the other road user;
b. the top-down vs. bottom-up allocation of the independent variable; and
c. the experimental implementation of the research question (see Figure 11.1).

In connection to the summary of empirical results and the methods' discussion we will conclude with an outlook for further research needs and crucial questions for future studies.

Detecting Motorcycles: What Did the Case Studies Tell Us?

The results suggest that PTWs' visual saliency influences their detection by other road users. That general result might sound to some extent as self-evident but – as we learnt in Chapter 3 – in many cases high visual saliency is neither the sufficient condition for detection as one might expect nor the sole factor that determines attention-capturing. However, several indicators in eye-tracking data reported within this volume hint at an influence of low-level visual characteristics of PTWs on the gaze behaviour of other road users. For example, researchers found greater saccadic movements preceding a fixation on a high saliency PTW compared to a low saliency PTW (see Chapter 4), shorter latency times to the first fixation on the PTW when it was presented in high saliency condition (see Chapter 6) and the studies consistently found longer fixation durations while inspecting a PTW when it was presented under low visual saliency condition relative to high saliency condition (see Chapters 4 and 6). All of these results gained from the eye tracking analyses indicate that, at an early stage of road-way scene inspection by road users, visual low-level features of PTWs guide road users' visual attention, and visually salient PTWs are more readily detectable and identifiable.

These findings are supported by further measures such as the detection distance for PTWs depending on their contrast to the background, corresponding with larger detection distances for higher object-background-contrasts or such as the detection rate for PTWs in tachistoscopic presentations (Chapter 7 and 8). These results are also strongly consistent with further findings on higher detection rates for PTWs with visual saliency enhancing light configurations or higher detection rates for PTW riders wearing clothes that make them visually distinguishable from the background (Chapter 10). So, the picture that emerges from the case studies provides some confidence with respect to the (detection related) efficiency of conspicuity treatments which aim for intensifying the visual salience of the PTW and/or its rider. In summary, increasing PTW saliency (with colour, lights, clothing and so on) shows positive impacts on detection in the laboratory. However, from a practical point of view it has to be noted that background factors may temper any benefits. Results from the case studies (for example Gershon and Shinar, Chapter 10) are hinting also to the fact that the visual prominence of a PTW is not only determined by the conspicuity treatment itself but is also crucially influenced by the present background given through the actual situation. Riding a bike is a dynamic activity – also with respect to the ever-changing (traffic) situation. Thus, dynamical changes of environmental background characteristics could for instance annul the advantages of bright rider clothing in a given situation whereas dark clothes might become more appropriate in order to enhance the visual prominence in this situation.

Before discussing the main results at the detection stage with respect to top-down influences, it seems worthwhile to recall the different ways this cognitive aspect (in contrast to physical aspects) has been considered within the case studies. One part of the studies used static group comparisons to focus on cognitive conspicuity. That is, the studies investigated differences between predetermined groups: car drivers and dual drivers (drivers who drive a car and also ride a bike) with the implicit assumption that dual drivers – compared to car drivers – dispose of different expectations and knowledge about motorcycles due to their experience with bikes (for methodological considerations, see below). The experimental manipulation of the prevalence of PTWs in driving simulation was another way which has been used in order to explore top-down effects on the detection performance (see Beanland et al., Chapter 9), and finally, the effect of expectations by other road users on detection can be examined through different variations of subject's task during an experimental series, that is, in terms of using a search vs. attention paradigm (see Gershon and Shinar, Chapter 10).

Concerning differences between car drivers and dual drivers, the studies surprisingly provide somewhat mixed results. With respect to the detection stage, the studies discovered fewer significant differences between the two groups than might be expected from the literature. Furthermore, differences seem to be restricted to particular driver tasks, for example dual drivers were found to detect PTWs farther away than car drivers when the PTW was depicted in the rearview mirror but not when the PTW was approaching the driver from the front (Rogé and Vienne, Chapter 8). One possible explanation for these inconsistent findings could be due to different demand levels of the tasks in the way that differences by experiences become more obvious for more demanding tasks such as the detection of PTW in the rear mirror whereas the differences diminish for less demanding tasks (floor effect). Findings from Salmon et al. (2014) provide further indications on why differences in detection performance between car drivers and riders may be affected by the location of the approaching PTW. In an on-road study, Salmon et al. examined differences in the situational awareness for different road user groups and showed that riders were more aware about the surrounding (including looking the side and behind) than car drivers. The results reported by Rogé and Vienne (Chapter 8 in the present volume) are furthermore accompanied by eye tracking data suggesting that dual drivers, based on their prototypical representation, indeed develop more efficient search strategies for motorcycles than car drivers: along with the finding of better detection of motorcycles in the mirror, it was found that dual drivers carried out more saccadic movements, showed shorter fixation durations and explored the visual scene in the mirrors more quickly than car drivers. Interestingly, an effect on gaze parameters by the allocation to a road user group (dual vs. car drivers) has also been identified in the context of road-way decisions – exclusive for vehicles depicted in the mirror (Underwood et al., Chapter 4). Admittedly, the authors found longer fixations on PTWs for dual drivers compared to car drivers and reported about indications that these longer fixations might be attributable to more intensive information acquisition by dual riders about further characteristics of the PTW (for example power) in order the make appropriate decisions. This finding and the results reported above also point to the critical role of the subject's experimental task with respect to the outcomes of studies.

Strong evidences for the effect of expectations on detection performance have been received when subjects were alerted to the occurrence of PTW by the experimental task (search conspicuity). Compared to unalerted participants within an attention conspicuity paradigm, subjects under search task conditions scored significantly higher detection rates up to near complete detection, even under conditions where the visual properties of PTWs and traffic situation were rather challenging. Furthermore, these results go along with findings where – exclusively within a search task – expectations of subject have been manipulated in a more covert way due to variations of the target vehicles' prevalence. That is, even when subjects are actively looking for motorcycles, implicit expectations about the prevalence seem to guide subjects' attentive focus, corresponding with the detection of motorcycles at a farther distance and with earlier detection in situations with a high prevalence of motorcycles compared to low prevalence conditions. Collectively these results point to the strong role that subjective expectancies play in determining an individual's performance in PTW detection.

Decision Towards and Appraisal of Motorcycles

With respect to the decisions that road users make when approaching PTWs, the results from the case studies suggest that visual salience not only influences the detection phase but also influences subsequent decision-making. For example, for verbally stated decisions, high visually salient PTWs were found to receive more careful and cautious decision outcomes than less visually salient PTWs within right-of-way scenarios and lane-change scenarios, respectively. Furthermore, the (safer) decisions have been made faster by the road users when a high visually salient PTW was approaching. If one looks at the safety implications of these findings the other way around, the critical circumstances for less conspicuous PTWs will become very obvious in that road users tend to make both less safe and also delayed responses in these scenarios. Both aspects potentially lead to decreased safety margins, and, thus, will increase the probability of multi-vehicle crashes.

Other approaches which shed light on the effects of PTWs' visual conspicuity on decision-making behaviour considered critical time gaps, and, respectively, gap-acceptance rates within simulated turning scenarios. These simulation studies (Mitsopoulos-Rubens and Lenné, Chapter 5) revealed that, in the case of an intended turning manoeuvre, participants more likely rejected gaps to turn ahead of a motorcycle when the motorcycle was more conspicuous by the application of low-beam headlights compared to the headlight-off-condition, in particular during safety critical situations (short time gaps). This is in line with the results reported above on more careful verbally stated decisions towards conspicuous PTWs. Moreover, the finding is supported by further results from simulation studies (Rößger et al., Chapter 6) considering the time gaps between subjects and an approaching target vehicle at the moment the subjects indicated they would give up their intention to turn in front of the target. When the visual front of motorcycles (target) has been accentuated by additional frontal lighting they earlier decided to give up the manoeuvre, corresponding with greater time gaps between subject and target. The results however suggested that this effect might be restricted to some extent to situations with poor visibility conditions (for example during night or twilight conditions).

In the case studies presented within this volume, top-down influences on road users' decisions towards PTWs were primarily assessed by the consideration of differences between car drivers and dual drivers. And again, the observed differences between both road user groups were less apparent and less consistent than both results (as reported above) related to effects of visual saliency on decision-making but also with respect to the reported effects of cognitive conspicuity on detection. The results, however, indicated as expected a tendency of more cautious decisions towards motorcycles by dual drivers compared to car drivers when the subjects were asked for stated judgements about safe or unsafe roadways (see Underwood et al., Chapter 3). This more careful decision-making behaviour by dual drivers was not reflected by results from simulation studies focusing on decisions during turning manoeuvres (see Mitsopoulos-Rubens and Lenné, Chapter 5). The authors could not observe differences in the decisions for both groups, but, instead, identified differences between car drivers and dual drivers in the turn performance, subsequent to the decision. Here again dual drivers revealed a more cautious behavioural pattern indicated by choosing greater safety margins and more efficient completion of the intermediate phase of the turn. The lack of hypothesised effects referring to road users' gap acceptance could be caused by different reasons. In addition to those already discussed in the chapter, it is conceivable that with generally increased probability for the acceptance a presented time gap (that is, for medium and long time gaps) the effect of top-down influences is shifted away from a binary decision outcome (accept vs. reject) towards decisions on the selection of a specific behavioural pattern that follows the original decision. According to this attempt to explain, effects should be rather observable in differences in terms of behavioural adaptations between car drivers and dual drivers than in differences concerning a discrete choice with two alternatives. Furthermore, the lack of effects might be also related to methodical shortcomings in a way that the simulation does not sufficiently capture the cues and complexity of the real world, and, therefore, potential effects do not become reliably evident (for further discussion on methodical issues see below).

Another – more general – remark concerns the quasi-experimental nature of the car driver vs. dual driver comparison when assessing cognitive aspects, like road users' expectations and experiences, and is referring also to other presented studies which carried out this analysis in order to shed light on the top-down impacts. Strictly speaking from a methodical point of view, it is a comparison between (pre-)existing groups which might imply potential confounding effects between experimental variables and further (group specific) factors, and, thus, is restricted in terms of explanative power to some extent (see, *inter alia* Sarris, 1992). Before we focus on the methodological aspects of the case studies in more detail in the next section, it should be noted already at this point that the lack of effects might be also ascribable to the way top-down factors have been operationalized.

Methodological Considerations

The studies presented in this volume used different strategies to consider the influences of PTWs' visual and cognitive conspicuity on road users' attention-capturing and decision-making behaviour. With respect to a continuum of ecological validity (in terms of the representation of reality), one might generally distinguish between screen experiments, and here furthermore between experiments using static (for example still pictures) and dynamic (for example video streams) stimuli, and simulator studies. That differentiation might sound somewhat uncommon but it seems useful in order to reflect the various ways in order to: a) vary visual characteristics of objects and measure attention/decision-making in conspicuity experiments; and b) to reflect the critical points which are associated therewith. Figure 11.2 provides a schematic overview of the approaches implemented within

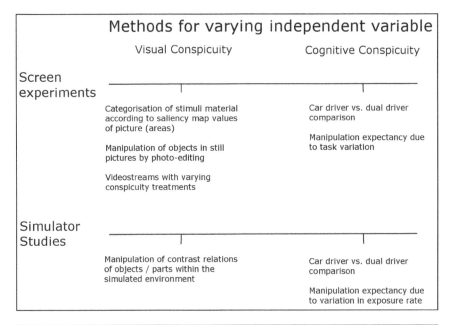

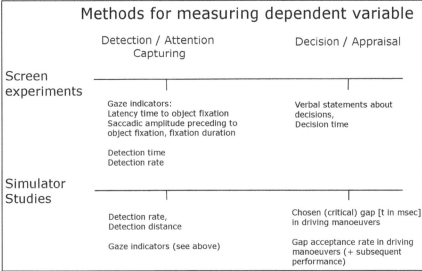

Figure 11.2   Schematic overview of methodological approaches

the case studies reported. It should be noted however that this is not a complete list of all possible approaches.

When we first look at the approach to focusing on visual conspicuity in screen experiments, two main critical points are obvious. First, due to the static characteristic referring to the presentation of the stimuli, there are essential determinants for attention-capturing are missing: looming or motion transients created by objects' dynamics (Franconeri and Simons, 2003). So, the question arises whether the static presentation of roadway scenes provides a sufficient approach in order to address attention-related research questions in the field of studying road users' behaviour? With respect to this question, Underwood et al. (Chapter 4 in this volume) reported that findings gained from experiments using still images are closely related to those which have been found from studies using dynamic presentation such as video streams. Furthermore, Scialfa et al. (2013; see also Scalfia et al., 2012) compared criterion-based validity of static vs. dynamic hazard perception tests and found that both instruments were equally capable of predicting self-reported errors and lapses, indicating a similar criterion-based validity. Though admittedly stimulated in another context, Stamps (2010) conducted an extensive meta-analysis focusing on the relations of subjects' responses towards simulated environments due to the use of static vs. dynamic media. His results showed strong overall correlations in subjects' responses between both kinds of presentation, suggesting that both presentation types produce equivalent results. Therefore, the author concluded that the choice of presentation should be rely on efficiency rather than concerns about validity (Stamps: 355).

According to their study outcomes, other authors (for example Koustanai et al., 2012), on the other hand, refer to weakness of a static presentation, especially when focusing on dynamic manoeuvres, such as overtaking, and encourage further research effort in order to shed light on differences in road users' detection abilities caused by various types of presentation. Consequently, and with reference to the validity reports cited above, it can be assumed that the experiments using static stimuli certainly have their justification in order to investigate attention and decision-making related behaviour of road users. With respect to the latter cited study, interpretation and generalizability of the results however should be carefully considered in each individual case and complementary approaches should take into consideration.

Compared to static representations, driving simulations provide a more immersive experience in a dynamic scenario. Such simulated scenarios allow researchers to purposefully exposure participants to traffic constellations which are known to have a higher risk and to measure participants' responses to specific manipulations in detail. However, there remain questions about the extent to which simulation can capture the complexity and variability of real-world performance. Concerning these questions, recent studies (Johnson et al., 2011) comparing simulated and on-road driving could show that the level of immersion of (fixed-base) simulator achieved relative validity as well as absolute validity, which was indicated by similar changes in physiological parameter (in response to surprising road events) and strong correlations between these parameters obtained under simulated and on-road driving conditions. With respect to visual attention, Konstantopoulos et al. (2010) showed that results on visual search strategies from driving simulator studies are comparable to those revealed from other environments, also suggesting that simulator studies provide a sufficient level of validity. Comparability between on-road and simulated studies has been also reported for hazard detection tasks and related gaze behaviour (Underwood et al., 2011).

A critical point refers to the presentation of conspicuity treatments, such as additional light sources, at computer screens but also in projections within driving simulations or similar. Due to the restricted luminance range, monitors or simulator screens cannot nearly provide a naturalistic representation of lights and the rendering of static and dynamic lights is clearly restricted by both software and hardware capabilities. The sensitivity of human photoreceptors ranges between 10^{-6} cd/m^2 and 10^8cd/m^2 whereas common displays are restricted to luminance values of 300 cd/m^2 (Ferwerda et al., 1996). In this context it might be interesting for experimenters that, under the keyword *perceptual realism*, Petit and Bremond (2010) developed a new method that allows the computation of high dynamic range (HDR) images based on the input from existing virtual environments and to render 8-bit images by a tone mapping operator, which can be then displayed at standard screens (as a 'perceptually realistic' low dynamic range (LDR) image).

Other authors of comparable studies (for example Hole and Tyrrell, 1995) consider the problem of the restricted luminance range less seriously when they argue that these restrictions should primarily diminish the differences between experimental and control conditions, and if at all, would therefore weaken the chance to reveal significant effects. If the studies despite this methodological restriction however are able to identify significant differences between experimental and control condition, one might assume that effects could be even stronger with real light sources. In conjunction to this argument it seems additionally valuable to examine the actual differences by manipulation checks in advance of an experimental series. So, for example, Mitsopolous-Rubens and Lenné (2012) empirically verified their distinction of two experimental conditions, in which the motorcycle headlight status has been varied, as low and high sensory conspicuity conditions due to applying the saliency algorithm developed by Itti and Koch (2000; see also Underwood, Chapter 3). Luminance measurements of the stimulus material provide another possibility in order to conduct a manipulation check and to be warranted that the variation of the independent variables is reasonable in experiments, and, thus, effective at least in a certain range. Luminance measurements imply the advantage that by using the photometric quantity in cd/m^2 they follow as closely as possible the human perception of brightness (Mather, 2006; Schlag et al., 2009).

When changing the focus to methodological approaches to examine effects of cognitive conspicuity, it has already been stated above that the usage of static group comparisons provides a common method but might make some critical assumptions that should be noted. Under the assumption of the cognitive conspicuity hypothesis the rationale beyond the comparison between predefined samples of riders and drivers is that the focused variables (for example such detection time) are affected by the differences in experiences with powered two-wheelers and/or different expectations related to this means of transport and so on. That rationale seems reasonable but it has to be taken into consideration that potential effects might also have been shown due to further background variables which systematically differ between rider and drivers. For instance, Horswill and Helman (2003) reported that the motorcycle population in the UK is systematically more male-dominated and younger than the car driver population. Furthermore, it seems plausible that the general choice for riding a bike is associated with specific personality traits of individuals which systematically differ from those who decide against a motorcycle as a means of transport. Although Horswill and Helman could not identify significant differences for matched samples of riders and drivers with respect to personality measures like sensation-seeking, other authors (Jackson and Wilson, 1993) reported for instance about higher sensation-seeking scores and higher impulsivity scores for motorcyclists when the results of the Eysenck Personality Profiler (EPP; Eysenck and Wilson, 1991) were compared to the population norm. With respect of systematic differences between the two groups, the question arises whether potential differences in attention indicators such as detection time are actually caused due to differences in cognitive conspicuity or whether differences are also explainable, for example by an association between sensation-seeking and focused-attention performance *per se* (Martin, 1985; Ball and Zuckerman, 1992) or by the link between age and detection performance (Wood, 1999; Chaparro et al. 2004).

In order to minimize such sources of secondary variance due to third variables, several strategies for studies on cognitive conspicuity are recommended: first, matched samples as considered by Horswill and Helman (2003) provide the opportunity to control the results for influences of third variables. Actually, these authors matched their groups of motorcyclists and non-motorcyclists according to age, gender, mileage, amount of driving/riding training and years of driving/riding licence validity, which seems a reasonable approach to achieve reliable and interpretable results. Second, obtaining additional control variables during a study as baseline measures will provide greater clarity regarding the true influence of factors under examination. In this context, the deliberate examination of alternative hypothesis offers an appropriate strategy in order to strengthen the validity of a study. Here, the work conducted by Rogé and Vienne (Chapter 8 in this volume) provides an excellent example. The authors examined in their study the detection performance of PTW by road users as a function of their riding experiences (dual driver vs. car drivers) under the assumption that the performance of dual drivers will be better because of their riding knowledge and more appropriate expectations concerning PTWs. Alternatively, they assumed that differences in the detection performance might be also occur because dual drivers, compared to car drivers, have an extended useful field of view, which generally enables them to process visual information in the periphery faster and easier. Their results do not support the assumption on the extended useful field of view but nevertheless they identified superior detection performance for dual drivers, which in turn provides further evidences for the cognitive conspicuity hypothesis. In addition to the mentioned approaches for controlling and/or examining the influence of third variables in the context of static group comparison, the conception and conduction of a *real* experiment is another option worth thinking about: here, comparability of the groups is warranted by randomization and the variation of the focused factors take place in fact through manipulation. This approach was applied by Beanland et al. (see Chapter 9 in this volume): in order to change subjects' expectations concerning motorcycles the authors systematically varied the subjects' preceding experience with motorcycles.

A further focus of the current chapter relies on the methods used to measure attention-capturing and decision-making. Depending on the situation dynamics, time-based or distance-based measures of object detection and detection frequencies traditionally have been considered as useful for, and commonly serve as object-related attention indicators. The link between attention and detection time seems very self-explanatory and obvious, but it should be noted that these measurements often require that the objects to be discovered by the subjects must be designated in advance, and in this case it is search conspicuity rather than the attention conspicuity that is determined (for differences between both types of conspicuity, see Gershon and Shinar, Chapter 10).

The measurement of eye movements in this context might provide a very useful method to counter this potential conflict as a highly direct measure of overt visual attention. Several authors furthermore point out that eye-tracking parameters are more closely related to visual attention than *key press behaviour* that occurs in the follow-up of response selection and skeletal muscle movement (for example Weierich et al., 2008; Armstrong and Olatunji, 2009). For instance, the time to first fixation on an object (Byrne et al., 1999) or the saccadic amplitude preceding to the fixation (Goldberg et al., 2002) represent eye-tracking metrics which have been proved in the literature to be suitable for measuring the attention-grabbing quality of objects. Regardless, the interpretation of eye-tracking metrics requires a careful consideration of the experimental circumstances. For instance, Crundall et al. (2012) identified longer fixation durations on motorcycles by dual drivers compared to car drivers while inspecting the road scenery at t-junctions, whereas according to results reported within this volume (Rogé and Vienne, Chapter 8), dual drivers showed shorter fixation durations than car drivers when the subjects had to detect motorcycles in the rearview mirror or motorcycles approaching in front of the subject. How can we explain such apparent contradictory findings? Longer fixations durations may indicate different underlying circumstances (Just and Carpenter, 1976): it indicates that the extracting of visual information is more difficult for the subject or, in the other way, it indicates that object is more engaging for the subject, respectively, the subject is more engaged in object's inspection. With respect to the differences in subjects' main tasks between both studies, which was a detection task in the study from Rogé and Vienne and a decision task about a roadway manoeuvre in the study conducted by Crundall et al., the contradiction seems more plausible. One might assume that because of their specific experience with and expectation about PTWs, dual drivers process more quickly the visual information which is required for PTW detection. On the other hand, if a decision is requested about roadway manoeuvres, dual drivers are more engaged in visual information intake related to PTW than car drivers, also because of their object-specific knowledge and specific object relevance. This explanation is supported in some way as Underwood et al. (Chapter 4) report about anecdotal results which suggest that dual riders more deliberately search for additional information (for example PTW power features) related to their decisions towards approaching PTWs and concurrently the authors reported that generally low saliency vehicles induced a greater inspection, indicated by longer fixation durations and a greater number of fixations.

The Way Forward

There can be little doubt from the research reviewed and presented in this book that conspicuity of PTWs influences how car drivers interact with PTWs in traffic. It is perhaps less clear how these laboratory and field study findings relate to crash risk, however, as noted in Chapter 2; perhaps improved methods of in-depth crash investigation will shed greater light on the link between these performance changes seen in the laboratory with real-world crash risk.

Results from the case studies presented in this book suggest that visual conspicuity treatments could offer some benefits in terms of better detection and better appraisal of PTW riders by other road users. However, as the results also indicate that potential benefits will be moderated by given environmental conditions (such as daytime, traffic density, surrounding background), the findings also encourage further types of research to resolve the visual conspicuity issues further. Naturalistic on-road studies with instrumented PTWs could provide a useful approach here to bridge the gap between findings from the laboratory research and the real-world conditions. Such studies would allow measuring the detection and appraisal of PTW under conditions of real-world complexity and with respect to the variability in road users' behaviour. Aside to the validation of results from the lab, on-road studies could provide furthermore necessary data for the evaluation of simulated environments. With respect to research on visual conspicuity in the lab, more realistic light rendering in simulations represents a remaining key challenge and despite promising approaches (for example Petit and Bremond, 2010) further data on the ecological validity of simulations is needed to ensure the appropriateness of this method.

Another challenge is the identification of conspicuity-related effects in crash data. At the moment, we can only conclude indirectly whether a conspicuity-related failure occurred. For example, for ROW crashes in left-turn scenarios where the car driver was at fault more likely suggest a conspicuity failure than a single vehicle PTW crash (De Craen et al., 2014). This of course remains a conclusion reached with a degree of uncertainty. Alternatively, we need to rely on witness reports and/ or what the driver said – and we know that this is not an objective indicator of a conspicuity-related failure. The progressive introduction of the connected vehicle concept where multiple sensors are fitting within vehicles (in-car but also fitted to road infrastructure and other road users) might also enable new possibilities to examine the role of visual conspicuity on crash occurrences more directly in the future. At least for research purposes – so irrespective from legal or acceptability questions – it is conceivable to continuously record the driver (and thus at least roughly his gaze behaviour) and the road scenario ahead for a certain time span (let's say 30 seconds), and in cases of (pre-)crash-related indications from the car sensor system, the records of the last 30 seconds can be transmitted to an external server for further in-depth analyses of the pre-crash situations, including driver's gaze behaviour related to visual aspects of the road way.

Related to the examination of cognitive conspicuity, the question arises about valid criteria to define driver and dual driver groups for between group comparisons. Further research is valuable which addresses issues on the type and the amount of experience, providing a reliable and valid basis to determine such groups as sensitive enough. In this context, for instance, a general research question could be: How much riding experience does a driver need to confer any potential benefits in terms of better cognitive conspicuity towards PTWs? Furthermore, the testing of the transferability and generalizability from findings gained in surveys and laboratory studies to real-world conditions in on-road studies represents another challenge in future research.

References

Armstrong, T. and Olatunji, B. (2009). What They See is What You Get: Eye Tracking of Attention in the Anxiety Disorders. *Psychological Science Agenda*, 23(3). Retrieved from: http://www.apa.org/science/about/psa/2009/03/science-briefs

Ball, S. and Zuckerman, M. (1992). Sensation seeking and selective attention: Focused and divided attention on a dichotic listening task. *Journal of Personality and Social Psychology*, 63(5), 825–31.

Byrne, M., Anderson, J.R., Douglas, S. and Matessa, M. (1999). Eye tracking the visual search of click down menus. *Proceedings of the CHI 99*.

Chaparro, A., Wood, J. and Carberry, T. (2004). Effects of Age and Auditory and Visual Dual-Tasks on Closed Road Driving Performance. *Proceedings of the Human Factors and Ergonomics Society 48th Annual Meeting*, 2319–22.

Crundall, D., Crundall, E., Clarke, D. and Shahar, A. (2012). Why do car driver fail to give way to motorcycles at t-junctions? *Accident Analysis & Prevention*, 88–96.

De Craen, S., Doumen, M.J.A. and Van Norden, Y. (2014). A different perspective on conspicuity related motorcycle crashes. *Accident Analysis & Prevention*, 63, 133–7.

Eysenck, H. and Wilson, G. (1991). *The Eysenck Personality Profiler*. London: Corporate Assesment Ltd.

Ferwerda, J., Pattanaik, S., Shirley, P. and Greenberg, D. (1996). A model of visual adaption for realistic image synthesis. *Proceedings of SIGGRAPH*, 249–58.

Franconeri, S. and Simons, D. (2003). Moving and looming stimuli capture attention. *Perception & Psychophysics*, 65(7), 999–1010.

Goldberg, H., Stimson, M., Lewenstein, M., Scott, N. and Wichansky, A. (2002). Eye tracking in Web search task: Design Implications. *Proceedings of Eye Tracking Research and Applications Symposium*.

Hole, G. and Tyrrell, L. (1995). The influence of perceptual 'set' on the detection of motorcyclists using daytime headlights. *Ergonomics*, 38(7), 1326–41.

Horswill, M. and Helman, S. (2003). A behavioral comparison between motorcyclists and a matched group of nin-mototcycling car drivers: factors influencing accident risk. *Accident Analysis and Prevention*, 35, 589–97.

Itti, L. and Koch, C. (2000). A saliency-based search mechanism for overt and covert shifts of visual attention. *Vision Research*, 40, 1489–506.

Jackson, C. and Wilson, G. (1993). Mad, bad or sad? The personality of bikers. *Personality and Individual Differences*, 14 (1), 241–2.

Johnson, M., Chahal, T., Stinchcombe, A., Mullen, N., Weaver, B. and Bedard, M. (2011). Physiological responses to simulated and on-road driving. *International Journal of Psychophysiology*, 81, 2013–208.

Just, M. and Carpenter, P. (1976). Eye fixations and cognitive processes. *Cognitive Psychology*, 8, 441–80.

Konstantopoulos, P., Chapman, P. and Crundall, D. (2010). Driver's visual attention as a function of driving experience and visibility. Using a driving simulator to explore drivers' eye movements in day, night and rain driving. *Accident Analysis & Prevention*, 42, 827–34.

Koustanai, A., Van Elslande, P. and Bastien, C. (2012). Use of change blindness to measure different abilities to detect relevant changes in natural driving scenes. *Transportation Reseach Part F*, 15, 233–42.

Martin, M. (1985). Individual differences in sensation seeking an attentional ability. *Personality and Individual Differences*, 6(5), 637–9.

Mather, G. (2006). *Foundations of Perception*. New York: Psychology Press.

Mitsopolous-Rubens, E. and Lenne, M. (2012). Issues in motorcycle sensory and cognitive conspicuity: The impact of motorcycle low-beam headlights and riding experience on drivers' decisions to turn across the path of a motorcycle. *Accident Analysis & Prevention*, 49, 86–95.

Petit, J. and Bremond, R. (2010). A high dynamic range rendering pipeline for interactice applications. *The Visual Computer*, 26, 533–42.

Salmon, P., Lenne, M.G., Walker, G.H., Stanton, N.A. and Filtness, A. (2014). Exploring schema-driven differences in situation awareness between road users: an on-road study of driver, cyclist and motorcyclist situation awareness. *Ergonomics*, 57(2), 191–209.

Sarris, V. (1992). *Methodologische Grundlagen der Experimentalpsychologie: Versuchsplanung und Stadien*. Munich: UTB.

Scalfia, C., Borkenhagen, D., Lyon, J., Deschenes, M., Horswill, M. and Wetton, M. (2012). The effects of driving experience on responses to a static hazard perception test. *Accident Analysis & Prevention*, 45, 547–53.

Schlag, B., Petermann, I. and Schulze, C. (2009). *Mehr Licht – mehr Sicht – mehr Sicherheit?* Wiesbaden: VS Research.

Scialfa, C., Borkenhagen, D., Lyon, J. and Deschenes, M. (2013). A comparison of static and dynaic hazard perception tests. *Accident Analysis & Prevention*, 51, 268–73.

Stamps, A. (2010). Use of static and dynamic media to simulate environments: a meta-analysis. *Perception and MotorSkills*, 111, 355–64.

Underwood, G., Crundall, D. and Chapman, P. (2011). Driving simulator validation with hazard perception. *Trasnportation Research Part F*, 14, 435–46.

Weierich, M., Treat, T. and Hollingworth, A. (2008). Theories and measurement of visual attentional processing in anxiety. *Cognition & Emotion*, 22, 985–1018.

Wood, J. (1999). How do visual status and age impact on driving performance as measured on a closed circuit driving track? *Opthalmic & Physiological Optics*, 1, 34–40.

Index